D0805611

Thin Layer
Chromatography

Analytical Chemistry by Open Learning

Titles in Series:

Samples and Standards
Sample Pretreatment
Classical Methods
Measurement, Statistics and Computation
Using Literature
Instrumentation
Chromatographic Separations
Gas Chromatography
High Performance Liquid Chromatography
Electrophoresis
Thin Layer Chromatography
Visible and Ultraviolet Spectroscopy
Fluorescence and Phosphorescence Spectroscopy
Infra Red Spectroscopy
Atomic Absorption and Emission Spectroscopy
Nuclear Magnetic Resonance Spectroscopy
X-Ray Methods
Mass Spectrometry
Scanning Electron Microscopy and X-Ray Microanalysis
Principles of Electroanalytical Methods
Potentiometry and Ion Selective Electrodes
Polarography and Other Voltammetric Methods
Radiochemical Methods
Clinical Specimens
Diagnostic Enzymology
Quantitative Bioassay
Assessment and Control of Biochemical Methods
Thermal Methods
Microprocessor Applications

Thin Layer Chromatography

Analytical Chemistry by Open Learning

Authors:
RICHARD J. HAMILTON
Liverpool Polytechnic, UK

SHIELA HAMILTON

Editor:
DAVID KEALEY

on behalf of ACOL

Published on behalf of ACOL, London
by
JOHN WILEY & SONS
Chichester · New York · Brisbane · Toronto · Singapore

Library of Congress Cataloging in Publication Data:

Hamilton, Richard J., 1936–
Thin layer chromatography.

(Analytical chemistry by open learning)
Bibliography: p.
1. Thin layer chromatography—Programmed instruction.
2. Chemistry, Analytic–Programmed instruction. I. Hamilton, Shiela, 1943–
II. Kealey, D. (David) III. Title. IV. Series.

QD79.C8H36 1987 543'.08956 86–32531
ISBN 0 471 91376 6
ISBN 0 471 91377 4 (pbk.)

British Library Cataloguing in Publication Data:

Hamilton, Richard J.
Thin layer chromatography. — (Analytical chemistry)
1. Thin layer chromatography
I. Title II. Hamilton, Shiela III. Kealey, D.
IV. ACOL V. Series
543'.08956 QD79.C8

ISBN 0 471 91376 6
ISBN 0 471 91377 4 (Pbk.)

Printed and bound in Great Britain

Analytical Chemistry

This series of texts is a result of an initiative by the Committee of Heads of Polytechnic Chemistry Departments in the United Kingdom. A project team based at Thames Polytechnic using funds available from the Manpower Services Commission 'Open Tech' Project have organised and managed the development of the material suitable for use by 'Distance Learners'. The contents of the various units have been identified, planned and written almost exclusively by groups of polytechnic staff, who are both expert in the subject area and are currently teaching in analytical chemistry.

The texts are for those interested in the basics of analytical chemistry and instrumental techniques who wish to study in a more flexible way than traditional institute attendance or to augment such attendance. A series of these units may be used by those undertaking courses leading to BTEC (levels IV and V), Royal Society of Chemistry (Certificates of Applied Chemistry) or other qualifications. The level is thus that of Senior Technician.

It is emphasised however that whilst the theoretical aspects of analytical chemistry can be studied in this way there is no substitute for the laboratory to learn the associated practical skills. In the U.K. there are nominated Polytechnics, Colleges and other Institutions who offer tutorial and practical support to achieve the practical objectives identified within each text. It is expected that many institutions worldwide will also provide such support.

The project will continue at Thames Polytechnic to support these 'Open Learning Texts', to continually refresh and update the material and to extend its coverage.

Further information about nominated support centres, the material or open learning techniques may be obtained from the project office at Thames Polytechnic, ACOL, Wellington St., Woolwich, London, SE18 6PF.

How to Use an Open Learning Text

Open learning texts are designed as a convenient and flexible way of studying for people who, for a variety of reasons cannot use conventional education courses. You will learn from this text the principles of one subject in Analytical Chemistry, but only by putting this knowledge into practice, under professional supervision, will you gain a full understanding of the analytical techniques described.

To achieve the full benefit from an open learning text you need to plan your place and time of study.

- Find the most suitable place to study where you can work without disturbance.

- If you have a tutor supervising your study discuss with him, or her, the date by which you should have completed this text.

- Some people study perfectly well in irregular bursts, however most students find that setting aside a certain number of hours each day is the most satisfactory method. It is for you to decide which pattern of study suits you best.

- If you decide to study for several hours at once, take short breaks of five or ten minutes every half hour or so. You will find that this method maintains a higher overall level of concentration.

Before you begin a detailed reading of the text, familiarise yourself with the general layout of the material. Have a look at the course contents list at the front of the book and flip through the pages to get a general impression of the way the subject is dealt with. You will find that there is space on the pages to make comments alongside the text as you study—your own notes for highlighting points that you feel are particularly important. Indicate in the margin the points you would like to discuss further with a tutor or fellow student. When you come to revise, these personal study notes will be very useful.

Π When you find a paragraph in the text marked with a symbol such as is shown here, this is where you get involved. At this point you are directed to do things: draw graphs, answer questions, perform calculations, etc. Do make an attempt at these activities. If necessary cover the succeeding response with a piece of paper until you are ready to read on. This is an opportunity for you to learn by participating in the subject and although the text continues by discussing your response, there is no better way to learn than by working things out for yourself.

We have introduced self assessment questions (SAQ) at appropriate places in the text. These SAQs provide for you a way of finding out if you understand what you have just been studying. There is space on the page for your answer and for any comments you want to add after reading the author's response. You will find the author's response to each SAQ at the end of the text. Compare what you have written with the response provided and read the discussion and advice.

At intervals in the text you will find a Summary and List of Objectives. The Summary will emphasise the important points covered by the material you have just read and the Objectives will give you a checklist of tasks you should then be able to achieve.

You can revise the Unit, perhaps for a formal examination, by re-reading the Summary and the Objectives, and by working through some of the SAQs. This should quickly alert you to areas of the text that need further study.

At the end of the book you will find for reference lists of commonly used scientific symbols and values, units of measurement and also a periodic table.

Contents

Study Guide

Thin-layer chromatography is one of the more popular of the separation techniques. We hope that you will find in this Unit sufficient theory to enable you to appreciate what is essentially a practical technique.

Because tlc is a practical science, you may find yourself forced to guess what the conditions will be to separate a new mixture of compounds. We would hope that we have provided such examples of classes of compound and background theory to enable you to make a very educated guess.

We have assumed that you have an understanding of chemistry equivalent to that of a student who has passed ONC (B.Tech.) in Physical Sciences. In particular you will need a knowledge of chemical equilibria and functional group chemistry. It is assumed that you have studied the ACOL Course: *Chromatographic Separations*, though we have presented some of the same material in a slightly different form.

If you already have some appreciation of tlc you may find that you can read Sections 1, 4 and 5 more quickly to leave you time to concentrate on Sections 2, 3, 6 and 7 and on the references given in the Bibliography.

In the Bibliography, we have listed some texts which will provide a number of separations additional to those which you have met in this Unit. In addition we have found C B F Rice and R Stock's book (now revised by Braithwaite and Smith) on *Chromatographic Methods*, and S G Perry, R Amos and P I Brewer's on *Practical Liquid Chromatography* to provide an alternative approach at about the same level as the Unit.

Practical Objectives

The following practical exercises would support:

– Part 2 (preparation of plates)

– Part 3 (choice of solvent)

– Part 5 (visualisation of separated spots)

– Part 6 (preparative tlc).

Simple practical exercises are *not* satisfactory for Parts 4 and 7.

Overall, 10–15 hours of practical work would be necessary to cover the practical objectives set out below. By the end of the practical course, the student should be able to:

– prepare a silica gel tlc plate;

– compare the performance of a self-prepared tlc plate with one which has been purchased pre-prepared;

– use the same sample mixture on the same adsorbent under different operating conditions to determine the reproducibility of the R_f values;

– select the appropriate solvent system for the separation of the known mixture;

– use tlc in a preparative mode to determine the recovery and purity of the separated compounds;

– separate a known mixture of compounds and visualise them by the three most popular visualising techniques.

Bibliography

1. A Braithwaite and F J Smith, *Chromatographic Methods*, 4th Edition, Chapman and Hall, London, 1985.

2. S G Perry, R Amos and P I Brewer, *Practical Liquid Chromatography*, Plenum-Rosetta, New York, 1972.

3. J C Touchstone and M F Dobbins, *Practice of Thin-Layer Chromatography*, J Wiley and Sons, New York, 1983.

4. J C Touchstone and D Rogers, *Thin-Layer Chromatography*, J Wiley and Sons, New York, 1980.

5. A Zlatkis and R E Kaiser, *High Performance Thin-Layer Chromatography*, Elsevier, Amsterdam, 1977.

6. *Thin-Layer Chromatography*, Ed. J G Kirchner and E S Perry in *Techniques in Chemistry* by A Weissberger, J Wiley and Sons, New York, 1978.

7. *Handbook of Chromatography*, Ed. G Zweig and J Sherma, C R C Press, Florida, 1972 onwards.

References 1 and 2 are books which deal with the subject at the same level as this Unit.

References 3, 4 and 5 are books which contain more extensive material at the same level as this Unit and in addition material which goes beyond this level.

References 6 and 7 are compendia of tlc analyses from which practical details may be taken.

Acknowledgements

Fig.2.2c is reproduced from the CAMAG catalogue: TLC 86 with permission from Camag.

We would like to thank Professor John Knox for the analogy of the Peruvian sand fly, Section 1.2. This description was first used in Edinburgh University Student Chemical Magazine.

We would like to thank B. J. Hamilton for the drawing of the sand fly on page 6.

1. Introduction

1.1. HISTORICAL SKETCH

Chromatography as a practical technique has developed from the Chemist's interest in being able to separate a mixture of compounds into it's constituents, with the ultimate aim of being able to identify these individual constituents. The earliest Chemists, the Alchemists, had theories about the separation of the early elements as described in an ancient record:

> You will separate the earth from fire, the subtle from the compact, gently with great skill.
>
> Thus you will possess the glory of the whole world and all obscurity will flee from you.'

We wish that we could guarantee such a pay-off these days, but we do hope that this Unit will remove some obscurity, at least, in the area of one separation technique – *Thin-Layer Chromatography* or tlc as it is usually called.

We'll leave you to decide.

Chromatography developed mainly in the 20th Century, although the earliest reference to a separation of coloured dyes on paper goes back to 1850. Runge applied a mixture of dyes to blotting paper and he recorded how the dye mixture separated into its constituent colours. This must be one of the most repeated experiments as numerous school pupils, including us and probably you, have doodled on blotting paper with a fountain pen, and noticed the separation of coloured inks. Even black biro ink can be persuaded to separate into its constituents with the addition of a drop of an organic solvent.

The name chromatography was first used in 1906 by a botanist called Tswett who worked to separate coloured plant pigments. He used two Greek words 'chromatos' – colour and 'grapha' to draw and despite the fact that the technique is no longer limited to coloured substances, the name has stuck. Tswett achieved separation by allowing a solution of mixed pigments to pass through a column of crushed chalk and he was able to see the separated compounds on his column.

Following column chromatography came the development, through the 1940's, of paper chromatography and, of principal interest here, the first description of thin-layer chromatography as described by two Russian workers Ismailov and Schraiber in 1938. Tlc was not developed initially as much as paper chromatography due to the easier control of the paper quality but tlc was given renewed impetus in the early 1950's by Kirchner and his group of co-workers. Finally, it was put on the map as a reliable separation technique by the work of Stahl in 1956. It was Stahl's attention to the details of the experimental technique that provided the necessary boost and meant that reproducible results did not lie in the lap of the Gods. Stahl recognised that the preparation of the plate, the solvent used, the shape of the tank and the overall conditions, eg humidity and temperature, could greatly influence the results of the experiment. He also noted that for consistency of results, conditions had to be controlled. These are all the factors which you will learn about as you work through this Unit.

So to the 1980's when tlc is a reliable, well documented and widely used separation technique with a sound basis of theory.

1.2. GENERAL DESCRIPTION OF tlc

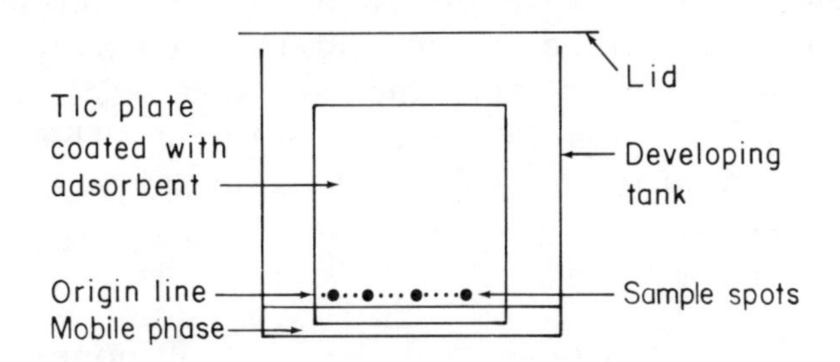

Tlc, like all analytical techniques, has its specialist vocabulary which you need to learn before you can understand the description of a tlc system.

The mixture of compounds to be separated is called the *sample* and the individual constituents are called the *components* or *solutes*. The sample, in solution, is 'applied' or 'spotted' on to a tlc *plate*, the plate consisting of a solid support, eg glass, plastic or aluminium 'coated' with an *adsorbent* layer or *stationary phase* specially chosen to effect the separation.

The 'loaded' plate is then placed in a *tank* containing an *eluting solvent* or *mobile phase* which will 'flow over' the plate. The solute must be applied at a measured distance from the bottom of the plate, called the *origin*.

Π Would you design your system such that the solute is placed on an origin which is:

(*a*) randomly chosen with respect to the surface of the eluting solvent,

(*b*) below the level of the surface of the eluting solvent,

(*c*) just above the level of the surface of the eluting solvent,

(*d*) towards the middle of the plate?

Remembering that to achieve separation the solute must, to some extent, be soluble in the eluting solvent otherwise it would not move at all, it is important that the origin is above the solvent surface to prevent the solute simply dissolving off the plate into the solvent, and, in order to allow sufficient distance on the plate for the separation to occur, you should choose the origin just above the surface. The point which the solvent reaches when it flows up the plate is called the *solvent front*.

Look at the separation of the mixture of acetophenone $C_6H_5COCH_3$ and hexadecanol $C_{16}H_{33}OH$ as an example.

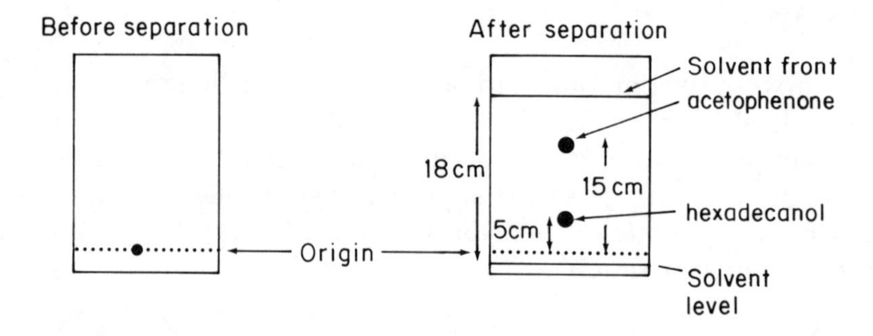

After separation, the mixture is divided into two separate constituents and these are identified by removing the plate from the tank, allowing any excess solvent to evaporate and, for this particular example, placing the plate in an iodine bath to 'colour' the spots.

The distance that the spots move up the plate is measured and quoted in terms of their R_f value where

$$R_f = \frac{\text{Distance moved by spot}}{\text{Distance moved by solvent front}}.$$

Π Calculate the R_f value of

(1) acetophenone,

(2) hexadecanol.

$$R_f \text{ value of acetophenone is } \frac{15 \text{ cm}}{18 \text{ cm}} = 0.83$$

$$R_f \text{ value of hexadecanol is } \frac{15 \text{ cm}}{18 \text{ cm}} = 0.28$$

It is clear that the R_f values are quite different and in order to explain this we need to look at what is happening in the system. We started with a mixture of two compounds, acetophenone and hexadecanol which were initially placed on the origin, in close contact with the stationary phase.

Π If the mobile phase is chosen so that both compounds have a greater affinity for the mobile phase than for the stationary phase would you expect:

(*a*) movement up the plate,

(*b*) no movement up the plate?

Clearly you would expect movement up the plate.

Π To explain the greater movement of the acetophenone compared with hexadecanol would you need to assume:

(*a*) that acetophenone has a greater affinity for the mobile phase than hexadecanol,

or

(*b*) that acetophenone has an equal affinity for the mobile phase,

or

(*c*) that acetophenone has a lesser affinity for the mobile phase than hexadecanol?

Since acetophenone travels further up the plate, it must have a greater affinity for the mobile phase than hexadecanol.

From an overall point of view, it looks as though this *differential migration*, as it is called, is a smooth process, ie it looks as though a substance which has a great affinity for the mobile phase is removed from the stationary phase and deposited near the solvent front whereas one with a lower affinity is deposited nearer the origin. However, from an individual molecule's point of view, the migration is anything but smooth. Consider the following useful analogy. An individual molecule can be likened to a mythical Peruvian sand fly. This fly lives on a very flat plateau in the Andes and it is unusual in not being able to fly. All the fly can do is jump up and down.

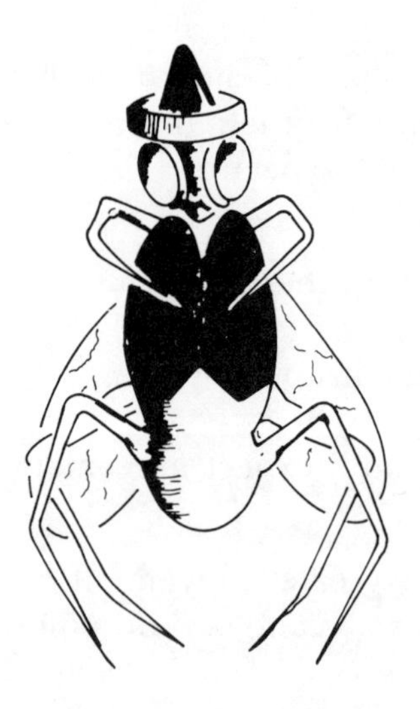

Peruvian sand fly

Peruvian sand flies are unusual in another way in that the males can jump twice as high as the females (or the females can jump twice as high as the males, it is equally frustrating). Finally, to set the picture, a constant wind blows over the plateau with constant velocity, v. So, picture the scene as the sun rises and the sand flies awake and start to jump up and down into the air.

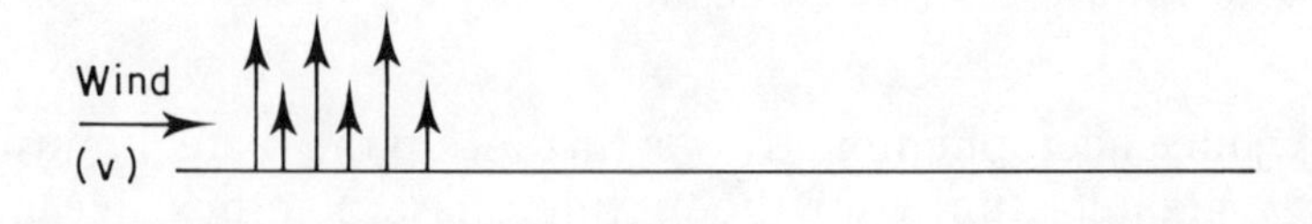

Initially, males and females are clumped together. After a finite time the high jumpers who spend longer off the ground are carried a greater distance than the low jumpers and the picture is as below.

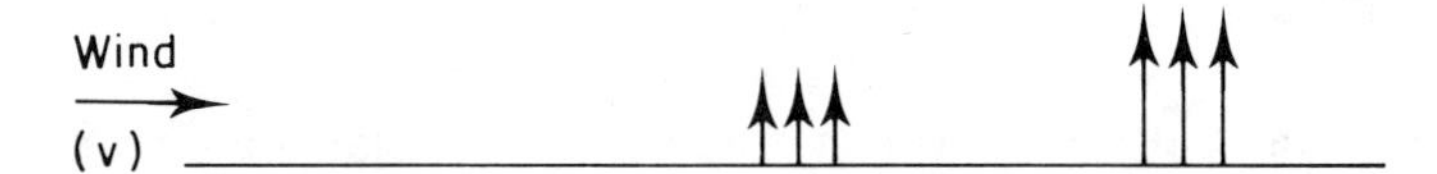

You can see why we describe the Peruvian sand fly as a mythical beast.

You may be forgetting, by this time, where the molecules come in, but the molecules (like the flies) leap backwards and forwards between the stationary phase (the ground) and the mobile phase (the air). Different molecules separate because they have different affinities for the mobile phase, but they execute a constant leap-frogging between the two, and so an equilibrium is set up.

The above solute-solvent equilibrium is only one of the ten possible interactions.

Π What other equilibria do you think could be set up in the solute, stationary phase, mobile phase chromatographic system?

$$\text{Solute} \underset{2}{\overset{1}{\rightleftharpoons}} \text{Solute} \underset{8}{\overset{7}{\rightleftharpoons}} \text{Solvent} \underset{10}{\overset{9}{\rightleftharpoons}} \text{Solvent}$$

$$\text{Solute} \underset{3}{\overset{4}{\rightleftharpoons}} \text{Adsorbent} \underset{6}{\overset{5}{\rightleftharpoons}} \text{Solvent}$$

1 and 2 correspond to the possibility of association/dissociation which occurs when there are two or more solute molecules in close contact with one another.

3 corresponds to the removal of or *desorption* of the solute molecules from the stationary phase.

4 corresponds to the attraction or *sorption* of solute molecules by the stationary phase.

5 corresponds to the *desorption* of solvent molecules from the stationary phase.

6 corresponds to the *sorption* of the solvent molecules by the stationary phase.

7 and 8 correspond to the association/dissociation between solute molecules and solvent molecules.

9 and 10 correspond to the possibility of association/dissociation which occurs when there are solvent molecules in close contact with one another.

To achieve a chromatographic separation we must have two simultaneously occurring mechanisms:

(*a*) A recurring dynamic equilibrium between sorption and desorption of the solute molecules represented by steps 3 and 4. This is the leap-frogging back and forth to the plateau that the Peruvian sand fly does.

(*b*) At the same time there must be competition between the solvent and the solute for the stationary phase.

When we use an adsorbent such as charcoal in the organic laboratory to clean up the products of a synthesis, we are not using chromatography. This is because we do not have competition, we have an equilibrium which lies very far to the right side, *viz* the solute adheres to the charcoal. Indeed all the coloured compounds adhere to the adsorbent and can be removed by filtering off the spent charcoal.

SAQ 1.2a

Choose the best definition for chromatography from the following:

(*i*) Chromatography allows the separation of mixtures by elution with a liquid.

(*ii*) Chromatography requires a competitive equilibrium between the stationary phase, the mobile phase and the sample molecules.

(*iii*) Chromatography permits the preferential sorption of a mixture of components from a gas.

(*iv*) Chromatography is a technique which allows an irreversible attachment between the samples to be separated and an adsorbent.

Summary

Tlc involves the interactions between a mobile phase, a stationary phase and the compounds, we wish to separate. There must be a dynamic equilibrium between sorption and desorption of the compounds and the stationary phase. In addition there must be competition between the solvent molecules and the solute molecules for the active sites in the stationary phase.

Objectives

You should now be able to:

- recognise the terms, adsorbent, solvent front, origin and mobile phase and give a definition;
- calculate the R_f value of any component from a separated mixture on the tlc plate.

2. Stationary Phase

2.1. ADSORBENTS

The solid layer supported on a non-porous plate in tlc is generally called the adsorbent although other stationary phases may be used in tlc that do not involve adsorption as the primary or only sorption mechanism. The nature and properties of the adsorbent are of crucial importance to the technique.

You may have asked yourself why ADsorbent and not ABsorbent, and it is important to distinguish between the two processes of adsorption and absorption at this stage.

ABsorption refers to the process by which a solute or substance is taken up and held within another substance. A sponge ABsorbs water. What about ADsorption, the process that we are interested in?

Some examples may give you a clue:

Animal charcoal is an adsorbent used in the organic chemistry laboratory to remove coloured impurities from a reaction product before recrystallisation in the purification process.

Charcoal was used in the gas masks issued during the 2nd World War with the hope that poisonous gases would be adsorbed on to the surface of the charcoal. Thus they could be removed from the air before it was breathed in.

Charcoal can also be used to remove excess chlorine gas from water purified by chlorination.

Adsorption is the ability of a solid to attract other molecules to its *surface* and to hold them at the surface. Thus charcoal can adsorb coloured impurities, poisonous gases and chlorine.

Adsorbents do not belong to a definite chemical class but they come with a wide variety of chemical structures which suggests that adsorption may be primarily a physical rather than a chemical process. As it is important in tlc to release the adsorbed substance, it should be noted that *no* chemical reaction has occurred between adsorbent and adsorbed materials.

However, adsorbents do *not* have particles like tiny billiard balls and a better picture shows that they have porous surfaces. The internal surfaces of the pores provide additional surface area (Fig. 2.1a). for adsorption.

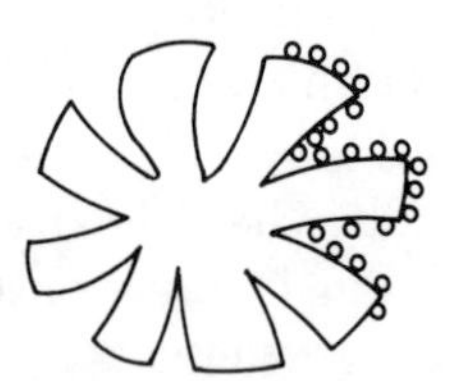

Fig. 2.1a. *Spherical adsorbent particle with pores and adsorbed molecules*

The cross-section of the silica particle is sometimes represented as in Fig. 2.1b which shows the inter-connecting pore structure.

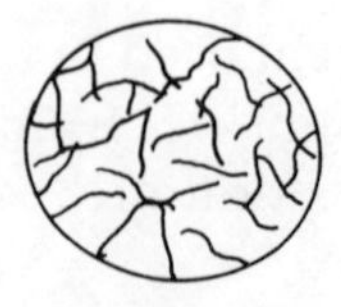

Fig. 2.1b. *Spherical adsorbent particle with inter-connecting pores*

The activity of an *adsorbent* is determined by its *surface area*, its *chemical nature* and the *geometrical arrangement* of atoms which make up its surface.

Hundreds of materials have been used as adsorbents – indeed the discoverer of classical column chromatography, Tswett, used almost 100. Fig. 2.1c lists a selection of adsorbents according to their activity and their surface area/unit mass. The activity depends on the production of 'active sites' on the surface of the adsorbent when it is prepared.

WEAK	MEDIUM	STRONG
Sucrose	Calcium Carbonate	Activated Magnesium Silicate
Starch	Calcium Phosphate	Activated alumina
Inulin	Magnesium Phosphate	Activated Charcoal
Talc	Magnesia	Fuller's Earth
Sodium Carbonate	Calcium Hydroxide	Silica Gel

Fig. 2.1c. *A selection of adsorbents*

The surface area of a medium adsorbent will be of the order of 10–50 metres per gram whereas a strong adsorbent will have an area of 100–500 metres per gram.

The following are some of the situations in which an analyst may find himself or herself.

(1) A research worker was trying to analyse aqueous solutions of potassium, sodium and calcium ions by tlc. He decided to try sucrose as an adsorbent with aqueous hydrochloric acid as the eluent.

Π Can you see an obvious disadvantage in this system?

Sucrose or sugar is soluble in water so that it would dissolve in this system.

In general, the adsorbent material must *not* dissolve in the chromatographic solvent.

(2) During the separation of steroids on basic alumina with a solvent of acetone, a research worker put 10 mg of the steroid mixture on to the tlc plate and recovered 15 mg of material.

Π How can you start with 10 mg of steroids and finish with 15 mg?

Clearly you *cannot* create matter so you must finish with not more than 10 mg of steroids. However, in the presence of the basic alumina, the following reaction can occur:

$$CH_3COCH_3 \xrightarrow{\text{Base}} CH_3\underset{\displaystyle CH_3}{\overset{\displaystyle OH}{\underset{|}{\overset{|}{C}}}}CH_2COCH_3$$

The additional mass came from this product which is called diacetone alcohol.

In general, the adsorbent material must *not* react with the solvent nor, as in the case above, cause dimerisation of the solvent.

(3) When a researcher tried to separate alkyl ethanoates from ketones on acidic silica gel, he obtained three spots on the tlc plate. Two spots corresponded to the unchanged alkyl ethanoates and the ketones. The third spot was identified as the hydrolysis product of the ethanoate, ie the alcohol formed by the acidic adsorbent.

In general, the adsorbent must *not* react with the solutes being separated.

(4) One researcher obtained the following results when he separated a three-component mixture on two plates of silica gel which he had prepared two weeks apart (Fig. 2.1d).

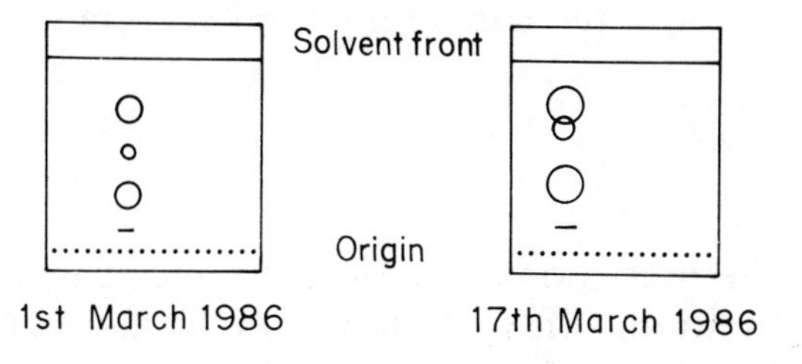

Fig. 2.1d. *Tlc separation on plates prepared two weeks apart*

It is obvious from these two plates that the researcher cannot depend on the separations that he is getting.

In general, an adsorbent must give reproducible results.

(5) A research worker tried to chromatograph a terpene alcohol on silica gel which he had activated by heating to 230 °C. He obtained a tlc plate which looked like the following:

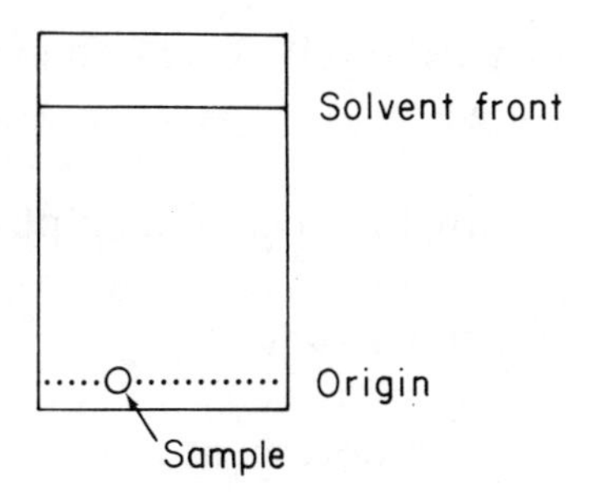

Fig. 2.1e. *Tlc plate after heating to 230 °C*

This is an example where the physical forces which hold the terpene to the surface are too strong.

In general, the adsorption process must be reversible.

Finally, the adsorbent must be reasonably inexpensive.

Now let us look at three commonly used adsorbents.

Silica Gel

Silica gel is the most popular adsorbent and it is prepared by the hydrolysis of sodium silicate followed by further condensation and polymerisation. The structure can be represented as follows:

```
          O-H          O-H          O-H
         /            /            /
 —O— Si —O — Si —O — Si —O—
      |        |        |
      O        O        O
      |        |        |
```

Its activity is due to the Si—OH (silanol) groups on the surface. The manufacturer controls the activity at the heating stage of the preparation. When used for tlc, the particle size of silica gel is in the range 5–10 μm mean diameter.

Some manufacturers use the following nomenclature to describe the various types of silica gel:

Silica gel G	with 13% calcium sulphate binder
Silica gel H	without binder
Silica gel F254	with fluorescent indicator
Silica gel UV254	with fluorescent indicator

Silica gel is slightly acidic in nature and we can use it to separate steroids, amino acids, alcohols, hydrocarbons, lipids, aflatoxins, bile acids, vitamins and alkaloids.

Alumina

The activity of silica gel depends on the number of Si—OH groups on the surface. For alumina (aluminium oxide), the activity depends on both the oxygen atoms and the aluminium atom, and the production methods are based on condensation of hydrated aluminium hydroxide.

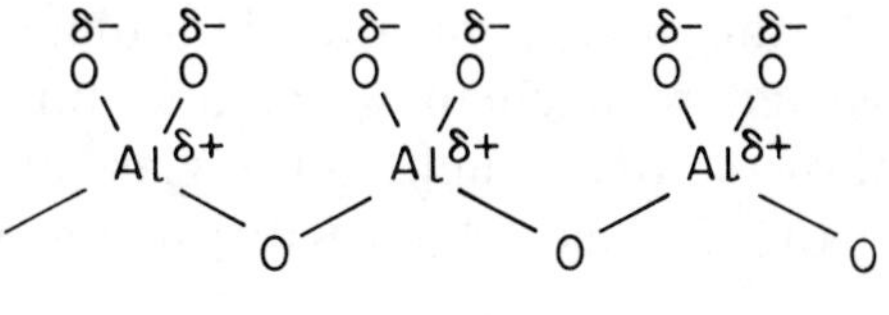

Alumina can be made with three degrees of surface acidity – acidic, neutral and basic, and the adsorbent is available with or without binder. Basic alumina is the most popular of the three. Alumina adsorbents can be used to separate sterols, dyestuffs, vitamins and alkaloids.

Cellulose

You may feel that it is unnecessary to make tlc plates coated with cellulose when paper could be used just as easily. In paper, the fibres leave gaps so that the eluting solvents flow along the fibres and fill the gaps with stagnant liquid. The solutes diffuse through these pools of liquid and so the spots tend to get larger as the elution proceeds. Tlc cellulose plates are made from small particles of cellulose, all of similar size, so that the solvent flows more evenly and the spots do not spread as much.

Cellulose is used to separate hydrophilic compounds such as sugars, amino acids, soluble inorganic ions and nucleic acids, which would adhere too strongly to alumina or silica.

We shall see later that with cellulose the sorption mechanism is predominantly partition where cellulose acts as a support for water adsorbed onto the surface.

2.2. PREPARATION OF PLATES

Earlier you will recall (Section 1.2) that we said that the adsorbent is in the form of a layer on a glass, plastic or aluminium plate.

2.2.1. Pre-treatment

Before we apply the adsorbent, the glass plate must be washed in soapy water, rinsed with clean water and finally with acetone to dry the plate thoroughly. Touching the glass after drying will leave fingerprints on the surface which will prevent the adsorbent from adhering.

2.2.2. Layer Thickness

For most qualitative uses, tlc requires a layer thickness of about 0.25 mm and as uniform as possible. However, when you wish to put a large sample (2–20 mg) onto a plate to enable you to isolate the separated spots (so-called preparative tlc), you need to have a thicker layer, 0.50–2.0 mm. If you put too much sample in one spot, overloading occurs and adjacent components will overlap with the result that the separated spots will be contaminated with other components.

We have commented on one major advantage of tlc, ie the ease with which it can be modified. One modification is found where the silica gel layer is *not* uniform. In Fig. 2.2a we can see the cross-section of a tlc plate where the layer is 1.0 mm thick at one side of the plate and reduces in thickness to 0.25 mm at the other side.

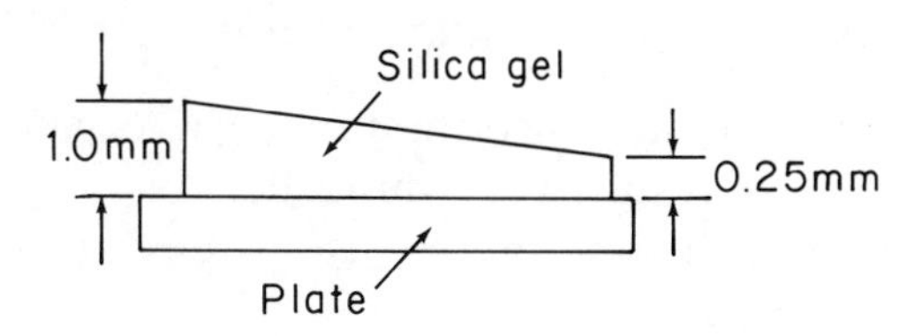

Fig. 2.2a. *Cross-section of a modified tlc plate*

These plates are very valuable when you need to chromatograph a large quantity of a mixture most of whose components you are *not* interested in. The thicker layer allows you to place a large amount of the mixture at one edge. As the solvent elutes the components you are especially interested in across the plate, the components reach the thinner part of the layer where they separate more efficiently. You need to adjust the solvent composition to ensure that the component that you are interested in has a high R_f value.

This is the situation that occurs in herbicide analysis where you have extracted a plant leaf with hexane to isolate a very small quantity of herbicide. Your extract will contain many other compounds in addition to the herbicide. A 0.25 mm layer gives best results for 5–25 μg quantities. In order to obtain 5 μg of the herbicide on the plate, you may need to apply at least 1 mg of the extract. Such a quantity would overload the 0.25 mm layer but it can be accepted by the thicker 1.0 mm layer of these modified plates.

We will now look at three ways of coating a glass plate.

2.2.3. Microscope Slides

The simplest way to make a cheap, useable tlc plate is to coat two microscope slides by holding them back to back and immersing them in a beaker which contains a slurry of silica in dichloromethane. As you withdraw the slides from the beaker, the dichloromethane evaporates off and you are left with two slides that are coated on one side only.

Π Can you see any disadvantage of this method?

The silica gel tends to settle out of the organic solvent very quickly. As a result, the whole beaker + slurry needs to be agitated each time before use to ensure that you have a uniform suspension.

These short tlc plates can be used without activation and they are very popular for rapid monitoring of reaction mixtures. This is mainly because they are cheaper than conventional 20 × 5 cm plates as they use much less silica gel.

2.2.4. Plates (20 × 20 cm)

For one plate you mix 5 g of silica gel with 15 cm^3 of water in a conical flask fitted with a ground glass stopper. The mixture is then shaken by hand till a smooth dispersion has been achieved. But you must ensure that this is achieved in 60 seconds after which time you pour the slurry onto the plate (Fig. 2.2b).

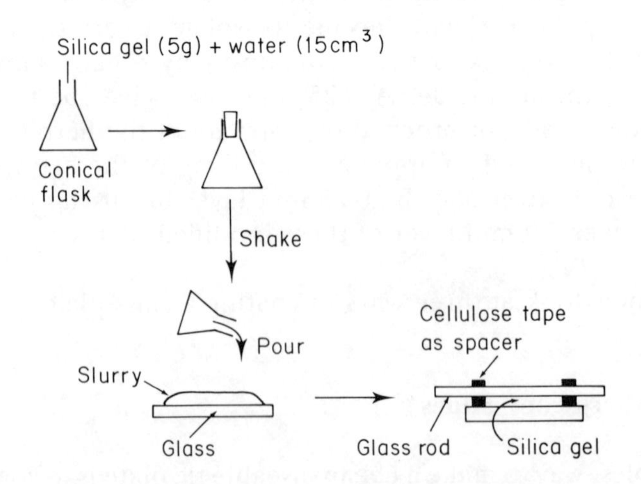

Fig. 2.2b. *Preparation of 20 × 20 cm plates*

Π Why must the slurry be poured within 60 seconds?

This short time is necessary because the binder tends to make the silica gel set quickly and so that it will not pour after a couple of minutes.

To form the layer using the least equipment, you spread the slurry by rolling a glass rod, which has 2 strips of cellulose tape around it to act as spacers, over the surface of the plate. The thickness of the layer is determined by the thickness of the cellulose tape, 0.25 mm being the normal thickness for qualitative separations (Fig. 2.2b).

It is often convenient to coat five plates (20 × 20 cm) at one time.

30 g of silica gel are slurried with 65 cm^3 of water and then poured into the reservoir of a commercial spreading system.

In one type (Fig. 2.2c) there is a fixed trough into which you pour the slurry of silica gel. You then move the plates by hand below the fixed trough from which the slurry flows to give a suitable layer thickness.

Fig. 2.2c *A commercial spreader*

Can you guess what the second type of spreading equipment will be? Yes, you have guessed it. Here, the trough can be moved and the glass plates are held on a plastic template. The trough is a rectangular box with neither top nor bottom. As the trough is pushed across the plates, the slurry flows out leaving a layer of the desired 0.25 mm thickness. For preparative tlc, the trough can be replaced with one which has a gap of 1 mm to provide a thicker layer which will allow more sample to be applied to the plate.

2.2.5. Activation

The plates are allowed to air dry for 15 minutes by which time the layer is sufficiently firm for you to transfer them to an oven set at 110 °C. Further drying at 110 °C for 30 minutes will produce a silica gel with a satisfactory activity.

It is possible by heating the plate to higher temperatures, eg up to 200 °C and for longer periods up to 4 hours, to produce a slightly more active silica gel. In general, it has been found that the thirty minutes at 110 °C regime gives a satisfactory compromise between activity and time spent in the preparation. In this stage of tlc as in all stages of the process, it is more important for you to keep the temperature and time constant, ie make the preparation as reproducible as possible. You must *not* make one batch at 110 °C and another batch at 200 °C.

2.2.6. Storage

When the 30 minutes drying/activation time is over, you must transfer the plates from the oven to a desiccator. It is easy to forget that silica gel is quickly de-activated by a moist atmosphere. However, commercial plates work perfectly well with no drying or storage. Have you ever used blue silica gel at home to keep double-glazed windows free from condensation? If you have, you will know that the silica gel needs re-activation very quickly in winter time. An activated silica gel plate will lose 50% of its activity in 3 minutes if it is placed in an atmosphere with a relative humidity of 50%. As a result, some companies insist that all handling of plates and the applying of mixtures to the silica gel should take place in a constant humidity room.

2.3. MODIFIED ADSORBENTS

As we commented earlier, one of the major advantages of tlc is that it can be modified easily, eg plates that incorporate silver nitrate into the silica gel can be made up in water containing 3.0 g of silver nitrate. When the plates have dried, we have a layer which contains silver ions. These can interact with the π-bonds of unsaturated molecules. The more π-bonds in a given molecule, the more strongly it will bind to the silver ions in the adsorbent. It is therefore possible to separate compounds according to the number of double bonds, eg a series of mono, di and trialkenes. Boric acid on silica gel can be prepared in a similar manner by slurrying the silica gel in a solution containing boric acid. This modified adsorbent is useful for separating diols and triols.

2.4. COMMERCIAL PRE-COATED PLATES

More and more laboratories now prefer to buy tlc plates already prepared. This is because the preparation of plates is time consuming and the apparatus expensive.

Moreover, manufacturers now supply a wide variety of pre-coated plates and the product is usually more consistent than that made in the laboratory.

The adsorbant can be supplied coated onto glass (thinner than that used for laboratory-prepared plates), aluminium or plastic (usually polyethylene terephthalate). Options usually include 'with binder' (G) or 'without binder' (H) and 'with fluorescent indicator' (F or F254) or without. The 254 signifies that maximum fluorescence will be observed at an excitation wavelength of 254 nm.

You can purchase plates of differing layer thickness ie 0.25 mm, 0.5 mm, 1.0 mm and 2.0 mm; 0.25 mm is the most widely used.

Each manufacturer will claim to have a standard product, ie a homogeneous coating, uniform thickness of layer, highly compacted adsorbent, firmly adherent coated layer, and consistent chromatographic characteristics. However, many researchers have shown that they achieved a particular separation with a plate from one manufacturer but *not* with the plates from another manufacturer. Thus selectivity and efficiency can vary significantly.

These ready coated plates can show some of the advantages claimed for self-prepared plates, in particular that they can be easily modified. One manufacturer supplies a plate with a 3 cm wide strip of kieselguhr attached along one edge, the remaining 17 cm strip having a silica gel coating. This pre-adsorbent zone of kieselguhr allows the solutes to be applied without the exaggerated care that is necessary to keep the spot area to a minimum which is needed for normal plates. Very little adsorption occurs in the kieselguhr layer so that the eluting solvent washes all the components up to the silica gel which they meet, therefore, as a compact band (Fig. 2.4a).

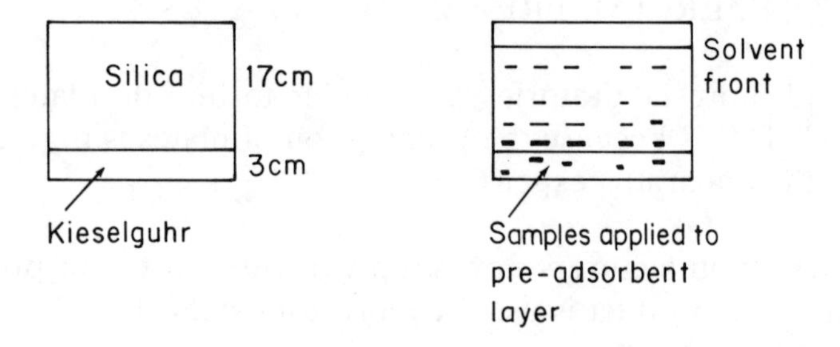

Fig. 2.4a. *Pre-adsorbent layer plate*

The pre-coated adsorbent can be modified by the manufacturer so that silver nitrate/silica plates for example are available commercially. Alternatively, some workers prefer to purchase silica gel plates and then coat the adsorbent layer by plunging them into a tray containing a mixture of ethanol : 20% (w/v) silver nitrate in water (1 : 1). After thirty seconds, the plate is taken out and the excess silver nitrate removed by drawing soft paper tissues over the surface. They are then activated by heating to 70 °C for 15 to 20 minutes.

SAQ 2.4a

Imagine that you are working in the County Analyst's Laboratory and your boss brings in a sample of methyl esters which he has prepared from the triglycerides present in a table margarine. He asks you to obtain 10 mg samples of both saturated and unsaturated methyl esters so that you can check that the label on the margarine packet is correct when it says 'High in Polyunsaturates'.

Which of the following tlc plates would you prepare?

(*i*) 0.25 mm layer thickness of silver nitrate/silica gel.

(*ii*) 1.00 mm layer thickness of silica gel.

(*iii*) 0.25 mm layer thickness of cellulose.

(*iv*) 1.00 mm layer thickness of silver nitrate/silica gel.

(*v*) 0.25 mm layer thickness of silica gel.

SAQ 2.4a

Summary

The activity of an adsorbent is dependent on the surface area of the particle and the surface area of the pores. The activity also depends on the chemical composition of the adsorbent. Silica-gel is the most popular adsorbent with alumina a poor second. Tlc plates can be prepared on almost any solid support (usually glass) ranging from microscope slides through 5 × 5 cm, 5 × 20 cm and 20 × 20 cm sizes. The adsorbent is added to the plate usually as a slurry, activated and then the prepared plate is stored in a dry place. Adsorbents can be modified as to thickness and to chemicals added to the adsorbent surface.

Objectives

You should now be able to:

- assess the properties needed for a substance to act as an adsorbent;
- prepare a silica-gel plate;
- recognise the benefits of some modified adsorbents;
- list the advantages of using a commercially pre-coated plate.

3. Mobile Phase

The three basic components of a chromatographic system are the adsorbent, the mobile phase and the sample.

Π Which of the above components do you, the analyst, have the most freedom to alter?

If you said sample, think again because, usually, you have no choice in this. Your company or the investigation will define what you are looking at.

If you said adsorbent, again, in a practical situation, you would want to start with the most easily available plates so you are limited to silica gel or alumina. You can find variations on these plates and you may eventually choose to change from adsorption to partition, but these are not likely to be the first changes you want to make.

So you are left with the choice of the mobile phase, and the secret to success in tlc depends, very often, on choosing the mobile phase correctly. You have an enormous choice of possible solvents, eg water, glycerine, ethoxyethane, ethanol.

So where do you start?

3.1. GENERAL PROPERTIES REQUIRED OF A MOBILE PHASE

Π Given the following list of properties, tick those that seem, at first sight, to be important in your choice of mobile phase.

1. Cheap	2. Expensive
3. Analar Grade	4. Reagent Grade
5. Low boiling	6. High boiling
7. Unreactive	8. Reactive

1. The first requirement of a solvent for a mobile phase is that it should be *cheap*! However, *not* cheap and nasty!! This may not be immediately obvious if you are setting up a small system on the bench but in extending your one-off analysis to a method which may be used in your company's laboratories many times a day, you will soon be in trouble with the company accountants if you need to order hundreds of litres of a very pricey solvent.

2. The solvent must be of high purity, so use the best available Analar Grade if possible. It is not much good trying to do a sophisticated qualitative separation if your mobile phase is going to vary from bottle to bottle because there may be varying amounts of impurities.

3. The mobile phase should be *unreactive* towards the solutes and the adsorbent. It would not be sensible to separate fatty acids with a mobile phase containing sodium hydroxide since the separation would be of the resulting salts – not of the fatty acids. A mobile phase which reacts with the adsorbent would clearly make interpretation of the results very difficult.

4. A *low boiling* solvent is generally preferable since the last step in the chromatographic process is to remove the plate from the tank and allow the mobile phase to evaporate. However, if it

is necessary to choose a *high boiling* solvent and you cannot achieve satisfactory results with a lower boiling one, the plate can be dried in an oven. (Remember that the oven must be vented to a suitable fume-hood).

Bearing in mind the general properties required of a mobile phase, we are still left with a very large choice of possible solvents. So how do you go about choosing a mobile phase for a particular analysis?

It can be seen that solvent selection has an element of black magic – 'the eye of newt and toe of frog' being replaced by 5% propanone and 3% ethoxyethane but in the next Section we hope that you will begin to appreciate the theory behind solvent selection and so remove the hit-or-miss element.

3.2. SOLVENT CHOICE

In tlc, the mobile phase has two vital jobs to do:

- It must displace the solute from the adsorbent so that the solute can be carried in the mobile phase across the plate.
- It must help to separate a mixture of solutes so that they can be deposited in different places and subsequently identified. This is termed solvent selectivity.

The effectiveness of a solvent in displacing solutes from an adsorbent is called its *eluting power*. In the early days of adsorption chromatography, a list of solvents was published whose eluting power for substances adsorbed on silica gel decreased in the order:

pure water, methanol, ethanol, propanol, propanone,
ethyl ethanoate, ethoxyethane, trichloromethane,
dichloromethane, benzene, methylbenzene, trichloroethene,
tetrachloromethane, cyclohexane, hexane.

SAQ 3.2a

In eluting samples on silica gel, is the first solvent of each pair more or less polar than the second?

Ethanol is more/less polar than benzene.

Trichloroethylene is more/less polar than cyclohexane.

Ethoxyethane is more/less polar than methanol.

Propanol is more/less polar than ethanol.

More recently, a function called the solvent strength parameter ϵ^{0} has been defined. This is the adsorption energy per unit area of standard adsorbent.

In general, the higher ϵ^{0} the more strongly the solvent interacts with the surface of the adsorbent and the more easily will the solute molecules be displaced leading to a *greater* R_f value.

The R_f values of a solute in each mobile phase will depend on the difference in their solvent strength parameters.

The solvent strength parameters, which can be measured, give rise to an 'eluotropic' series of solvents when the solvents are arranged in order of increasing ϵ^{0} (Fig. 3.2a).

Solvent	$\epsilon^o(Al_2O_3)$	Viscosity, cP, 20°	RI	UV cutoff, nm	Boiling point, °C
Fluoroalkanes	−0.25	–	1.25	–	–
Pentane	0.001	0.23	1.358	210	36
2,2,4-Trimethylpentane	0.01	0.54	1.404	210	118
Heptane	0.01	0.41	1.388	210	98.4
Decane	0.04	0.92	1.412	210	174
Cyclohexane	0.04	1.00	1.427	210	81
Cyclopentane	0.05	0.47	1.406	210	49.3
Carbon disulphide	0.15	0.37	1.626	380	45
Tetrachloromethane	0.18	0.97	1.466	265	76.7
1-Chloropentane	0.26	0.43	1.413	225	108.2
Diisopropyl ether	0.28	0.37	1.368	220	69
2-Chloropropane	0.29	0.33	1.378	225	34.8
Methyl benzene	0.29	0.59	1.496	285	110.6
1-Chloropropane	0.30	0.35	1.389	225	46.6
Chlorobenzene	0.30	0.80	1.525	280	132
Benzene	0.32	0.65	1.501	280	80.1
Bromoethane	0.37	0.41	1.424	225	38.4
Ethoxyethane	0.38	0.23	1.353	220	34.6
Trichloromethane	0.40	0.57	1.443	245	61.2
Dichloromethane	0.42	0.44	1.424	245	41
Tetrahydrofuran	0.45	0.55	1.408	220	65
1,2-Dichloroethane	0.49	0.79	1.445	230	84
Butanone	0.51	0.43	1.381	330	79.6
Propanone	0.56	0.32	1.359	330	56.2
Dioxane	0.56	1.54	1.422	220	104
Ethyl ethanoate	0.58	0.45	1.370	260	77.1
Methyl ethanoate	0.60	0.37	1.362	260	57
Pentan-1-ol	0.61	4.1	1.410	210	137.3
Dimethyl sulphoxide	0.62	2.24	1.478	270	190
Aniline	0.62	4.4	1.586	325	184
Nitromethane	0.64	0.67	1.394	380	100.8
Acetonitrile	0.65	0.37	1.344	210	80.1
Pyridine	0.71	0.94	1.510	305	115.5
Propan-2-ol	0.82	2.3	1.38	210	82.4
Ethanol	0.88	1.20	1.361	210	78.5
Methanol	0.95	0.60	1.329	210	65.0
Ethylene glycol	1.11	19.9	1.427	210	198
Ethanoic acid	Large	1.26	1.372	251	118.5

* ϵ^o is defined as zero on alumina when pentane is used as the solvent.

Fig. 3.2a. *An eluotropic series for polar adsorbents**

Π Indicate whether each of the following statements are true or false:

(*a*) ϵ^o can be defined as zero on silica where pentane is used as the solvent since it is defined as zero on alumina with this same solvent. TRUE / FALSE.

Statement (*a*) is false because when we change the adsorbent, the ϵ^o value will change too. The ϵ^o is defined in terms of solvent/adsorbent attraction. ϵ^o values on silica are 0.77 times the values on alumina.

(*b*) In using the eluotropic series (eg Fig. 3.2a) there will be one order of solvents listed in order of increasing ϵ^o for each adsorbent. TRUE / FALSE.

Statement (*b*) is true. There is an eluotropic series for each adsorbent.

The series for alumina and silica gel are very similar but both are very different from the eluotropic series for activated carbon. Going back to the title of this Section, 'Solvent Choice', knowledge of an eluotropic series gives us a starting point for a first trial with a mobile phase to determine which solvent is best.

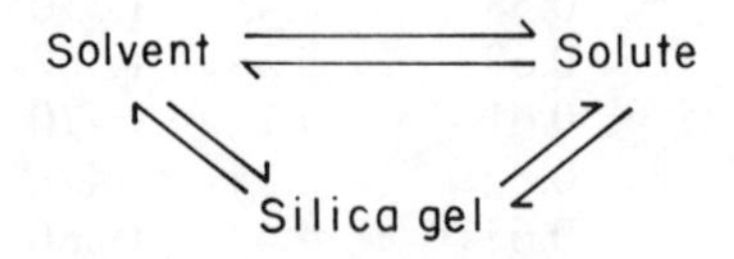

Remember that there is an equilibrium between solvent, solute and adsorbent, eg silica gel, where the solvent competes with the solute for the active sites on the silica gel.

In a first trial to decide which solvent to use for a particular sample, cyclohexane (ϵ^o = 0.04) was chosen. After application of the sample to the plate and development with cyclohexane, the plate looks as below:

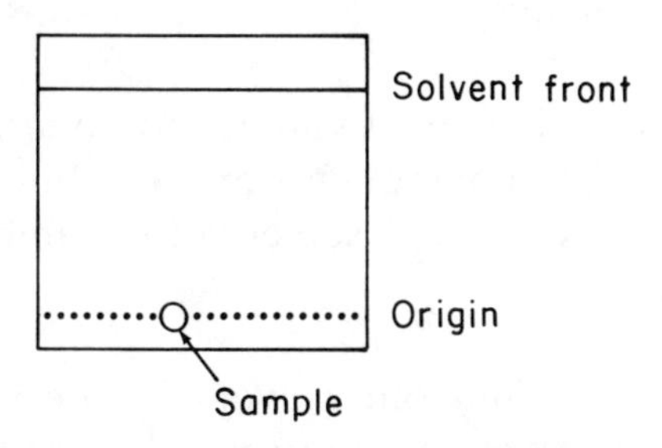

Π What do you deduce about the use of cyclohexane as a solvent?

The cyclohexane is too 'weak' a solvent to displace any of the sample components which are firmly adsorbed on the silica gel. Thus the next trial will require you to choose another solvent.

Π Would you choose pentane ($\epsilon^{0} = 0.00$) lying below cyclohexane in the eluotropic series?

No! The usefulness of the eluotropic series is that it warns you that, if cyclohexane $\epsilon^{0} = 0.04$ cannot move the sample, a solvent like pentane with an even lower value will be an even weaker mobile phase and will allow the sample to be more strongly adsorbed at the origin.

For a third trial, you may choose dichloromethane ($\epsilon^{0} = 0.42$) as the mobile phase and the separated plate might then look like the one shown below:

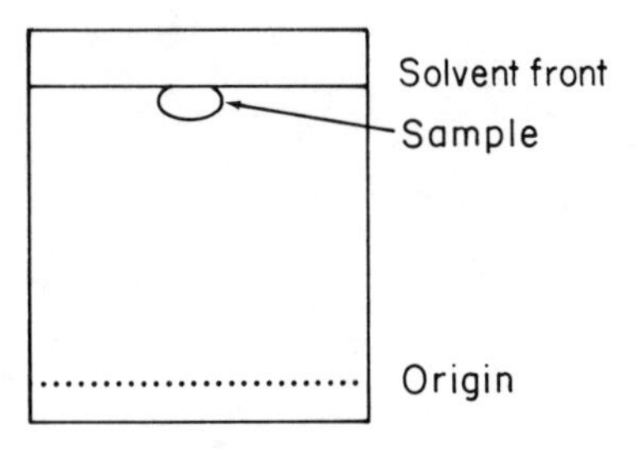

Π How would the eluotropic series help you to choose an alternative solvent?

In this case dichloromethane is too polar as seen by the fact that the sample components are all eluted with the solvent front, so the next time a solvent of ϵ^0 value between 0.04 and 0.42 is used.

It may be possible to find one solvent in the range 0.04 to 0.42, but the possibilities are greatly increased if mixtures of solvents are used, and ϵ^0 values can be tailored to the analytical requirements.

SAQ 3.2b

In the third trial above if you were trying to find a suitable mobile phase by mixing dichloromethane with another solvent, which of the following would you choose:

1. pentane

2. trichloromethane

3. ethanol?

Fig. 3.2b lists a number of solvent mixtures with their corresponding ϵ° values. An alternative way of presenting these figures is shown in Fig. 3.2c.

On silica:

Methanol : Ethoxyethane	Eluent Strength/ϵ°	Acetonitrile : Methanol
0.25 : 99.75	0.40	
0.75 : 99.25	0.45	
1.70 : 98.30	0.50	
3.50 : 96.50	0.55	
8.00 : 92.00	0.60	
18.00 : 82.00	0.65	70.0 : 30.0
42.00 : 58.00	0.70	60.0 : 40.0
100.00 : 0.00	0.73	0.0 : 100.0

On alumina:

2-Chloropropane : Pentane	Eluent Strength/ϵ°	Ethoxyethane : Pentane
8 : 92	0.05	4 : 96
19 : 81	0.10	9 : 91
34 : 66	0.15	15 : 85
52 : 48	0.20	25 : 75
77 : 23	0.25	38 : 62

Dichloromethane : Pentane	Eluent Strength/ϵ°	Ethoxyethane : Pentane
13 : 87	0.20	25 : 75
22 : 78	0.25	38 : 62
34 : 66	0.30	55 : 45
54 : 46	0.35	81 : 19

Fig. 3.2b. *Eluent strengths of binary mixtures*

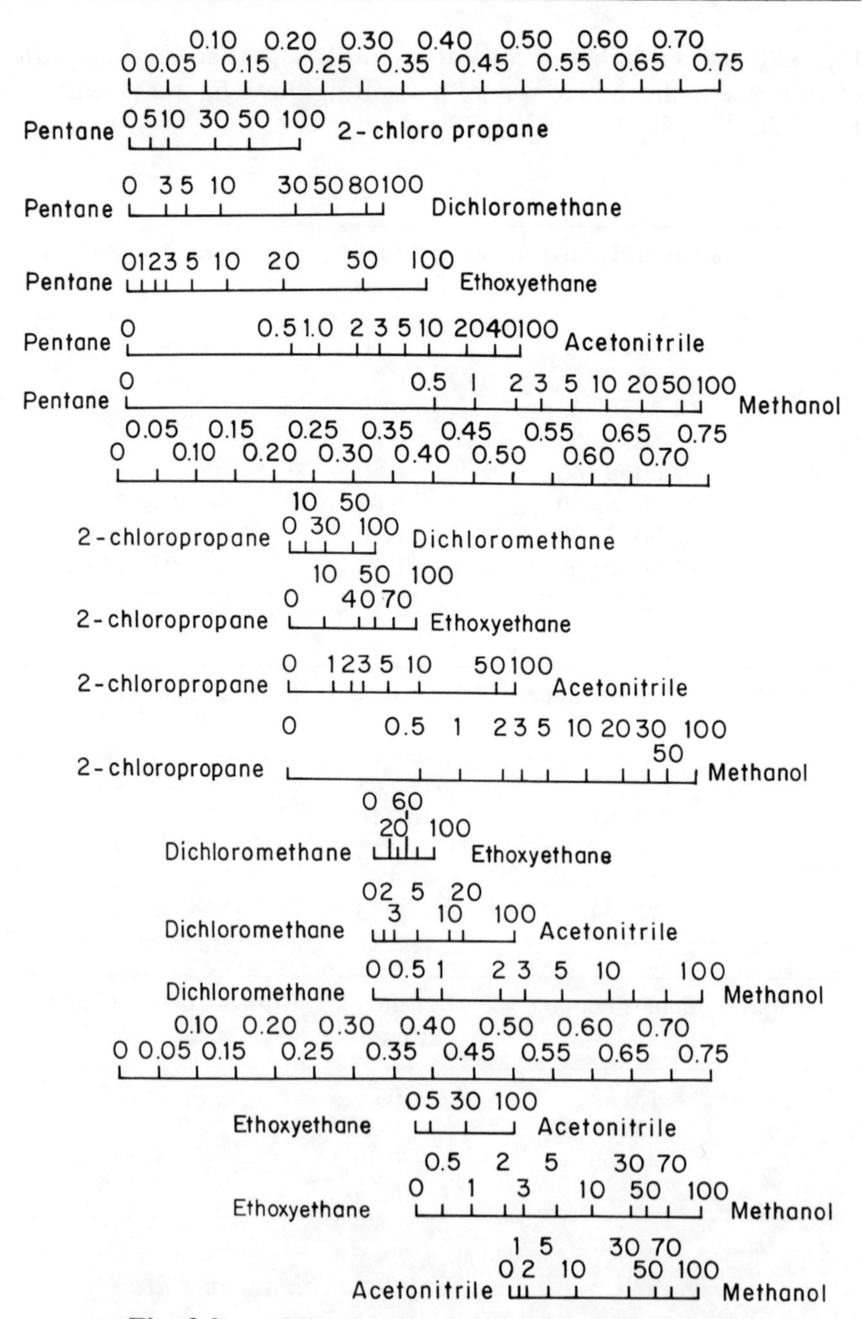

Fig. 3.2c. *Mixed solvent strengths (silica gel)*

The values in Fig. 3.2c allow you to determine the overall ϵ^0 value of binary mixtures of solvents. They are presented in the form of a nomograph. We have said that 100% pentane is defined as having an ϵ^0 value of 0.00. Thus in the first line of the nomograph, we see how increasing the proportion of 2-chloropropane causes the value to increase. Thus, at 50 : 50 2-chloropropane : pentane, the value is 0.16 and at 100% 2-chloropropane the value is 0.225.

The second line shows that small amounts of dichloromethane in pentane alter the ϵ^0 value very significantly, so that 10% dichloromethane in pentane gives a value of 0.12. The final line represents the amount of methanol in pentane where even 0.5% methanol corresponds to a value of 0.4.

You can see that, in general, to obtain a mixture of a particular ϵ^0 value, you need to drop a perpendicular line from the ϵ^0 values line to meet the appropriate binary mixture line.

SAQ 3.2c

Which of the following concentrations of acetonitrile in pentane would give a solvent strength (ϵ^0 value) of 0.45?

(*i*) 1.0%,

(*ii*) 30%,

(*iii*) 40%,

(*iv*) 80%.

SAQ 3.2d

Which of the following proportions of acetonitrile in dichloromethane would give a solvent strength of 0.35?

(*i*) 1.0%,

(*ii*) 3.0%,

(*iii*) 10%,

(*iv*) 100%.

SAQ 3.2e

What is the ϵ^0 value of a solution containing 10% methanol in ethoxyethane?

SAQ 3.2f

Two components A and B in a mixture have almost identical R_f values of 0.02 and 0.03 on silica gel with a mobile phase of 2-chloropropane : pentane = 8 : 92.

Which of the following mobile phases would you use to try to increase the separation of the components A and B?

(*i*) pentane,

(*ii*) methyl ethanoate,

(*iii*) acetonitrile : pentane 20 : 80,

(*iv*) methanol : dichloromethane 1.5 : 98.5.

A simple rule-of-thumb has been used for many years for unknown samples. You start with a very low polarity solvent and then add a more polar solvent in increasing steps to give mixtures containing respectively 2, 4, 8, 16 and 32%. Fig. 3.2c shows that each of these increases in % corresponds to an increase in eluent strength of about 0.05 unit.

To summarise if the R_f is too high, choose a mobile phase of lower ϵ^o value. If it is too low, go to a higher ϵ^o value, making fine adjustments by blending two appropriate solvents.

Summary

The chemical and physical properties of a solvent determine its suitability as a mobile phase. The relative efficacy of a solvent can be observed by examining its position in the eluotropic series. To determine the best solvent system, start with a non-polar solvent then increase the proportion of a more polar solvent in steps such that the ϵ^o value will increase by units of 0.05. The eluting power of a binary mixture of solvents can be determined by using the nomograph in Fig. 3.2c.

Objectives

You should now be able to:

- state the general properties needed for an eluting solvent;

- recognise the significance of Snyder's solvent strength parameter ϵ^o;

- given an eluotropic series, assess which solvent will be more effective as an eluting solvent;

- minimise the number of trial experiments you would perform to achieve some separation of a known mixture of compounds;

- calculate the ϵ^0 values of any binary mixture of solvents (pentane, 2-chloropentane, dichloromethane, ethoxyethane, acetonitrile and methanol).

4. Sample

This is the third major component in the chromatographic system.

What are the other two?

We are going to consider that these other two components, ie the adsorbent and the mobile phase, remain unchanged. Can we find any general rules about which sample component will run furthest up the plate and which will have the smallest R_f?

When we are introducing organic chemistry, we explain that two major concepts enabled chemists to simplify the classification of the thousands of organic molecules.

Π Do you recall these two simplifying concepts?

They are the idea of a homologous series and the effects of functional groups. Differences in the number and nature of functional groups can be used to explain variations in the R_f values of sample components.

A study of adsorption energies, Q_i^0, for a selection of compounds enables predictions as to chromatographic behaviour to be made. Fig. 4.1a lists functional groups attached to an aliphatic group R according to their adsorption energies on both silica gel and alumina. This is an approximate order in which compounds containing these functional groups would be eluted on tlc plates.

Group	Silica Gel Q_i^o	Interaction	Alumina Q_i^o	Interaction
R—CH_3	+0.07	Van der Waal	−0.03	Van der Waal
R—CH_2—	−0.05	Van der Waal	+0.02	Van der Waal
R—Cl	+1.32	Inductive	+1.82	Inductive
R—O—R	+3.61	Proton Acceptor	+3.50	
R—CHO	+4.97	Proton Acceptor	+4.73	
R—NO_2	+5.71	Inductive	+5.40	Inductive
R—CO_2R	+5.27	Proton Acceptor/ Inductive	+5.00	Inductive
R—COR	+5.27	Proton Acceptor/ Inductive	+5.00	Inductive
R—OH	+5.60	Hydrogen Bond	+6.50	
R—NH_2	+8.00	Hydrogen Bond	+6.24	
R—CO_2H	+7.60	Hydrogen Bond	+21.00	Chemisorption

Fig. 4.1a. *Adsorption energies for functional groups*

What forces cause the solute molecules to be adsorbed?

We can subdivide the possible interactions between the adsorbent and the solute into four groups:

Group 1. It is known that neutral molecules are held together by forces known as London Dispersion Forces (an alternative name is Van der Waal's forces.) Such forces are purely physical and no chemical bonds are formed. Because these physical adsorption forces allow the equilibria between adsorbent, solute and solvent (Section 1.2) to be set up easily, chromatographic separations are good and spots are symmetrical. In Fig. 4.1a you can see that the Q_i^o value for methylene groups is low so that members of a homologous series have nearly identical adsorption energies. Adsorption chromatography tends to separate according to the functional group, so compounds are separated by type, not by relative molecular mass.

Group 2. If the solute has a polar bond which has a permanent field associated with it, eg C—Cl or $C—NO_2$, the electrons in an adjacent molecule can be polarised so as to give rise to an induced dipole moment. Such forces appear to be more important in the case of alumina than silica. Fig. 4.1a shows that these groups have a larger Q_i^0 value than hydrocarbons and so molecules containing these groups have small R_f values. Again, because the forces are physical, symmetrical spots are formed.

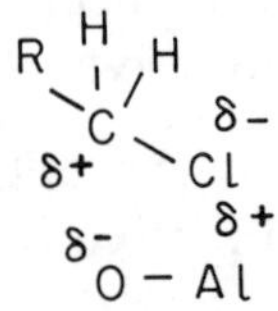

Fig. 4.1b. *Association between a chloroalkane and the surface of alumina*

Group 3. If an adsorbent has a nucleophilic polar surface, the solute molecules may be able to interact with the surface by hydrogen bonding, eg acids and alcohols behave in this way.

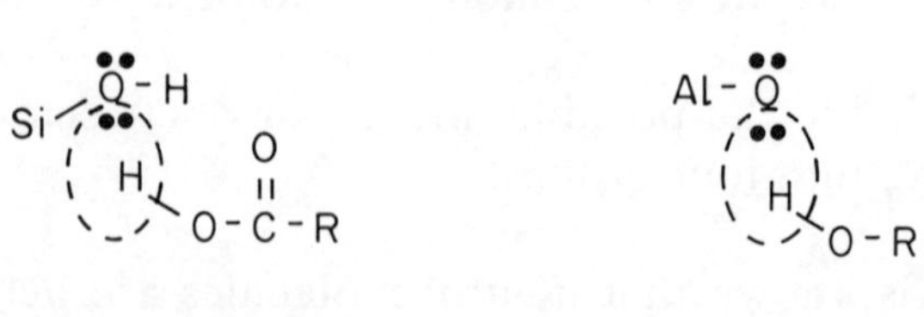

Fig. 4.1c. *Association between an acid and the surface of silica, and between an alcohol and the surface of alumina*

The process can be reversed with silica, which has a surface Si—OH group, provided that the solute molecule has a group which can act as a proton acceptor. Thus ethers, nitriles and aromatic rings can interact by this mechanism. This force is also a physical interaction but because it is so strong there is always a danger of the alcohols and acids appearing as spots with tails.

Fig. 4.1d. *Association between ethers and nitriles with the surface of silica gel*

Group 4. The last type of interaction is a chemical one rather than a physical one. With some adsorbents, a strong covalent bond will be formed with certain solutes. Alumina with its basic sites causes acids to bind by this interaction which is called chemisorption. These very strong chemical forces lead to poor separations because the spots may remain close to or at the origin of the plate.

You can see that the order of functional groups in Fig. 4.1a, like the eluotropic series for the mobile phase, enables us to predict an order of elution.

SAQ 4.1a

In relation to the separation of an aliphatic ketone from an aliphatic alcohol on silica gel, which one of the following paragraphs, (*i*)–(*iv*), is correct?

(*i*) The aliphatic alcohol R—OH will have an R_f value of 0.2 compared to a value of 0.5 for the aliphatic ketone R—CO—R because the alcohol interacts only by London Dispersion forces. ⟶

SAQ 4.1a (cont.)

(*ii*) The aliphatic ketone will have an R_f value of 0.2 compared to a value of 0.5 for the aliphatic alcohol because the ketone can interact by hydrogen bonding as a proton donor.

(*iii*) The aliphatic alcohol will have an R_f value of 0.2 compared to a value of 0.5 for the aliphatic ketone because the alcohol interacts by hydrogen bonding as a proton acceptor and as a proton donor.

(*iv*) The aliphatic ketone will have an R_f value of 0.2 compared to a value of 0.5 for the aliphatic alcohol because the ketone will interact by chemisorption.

Selective adsorption occurs when there is competition between the solutes and the mobile phase for the surface of the adsorbent. In general, polar compounds are more strongly adsorbed by polar solids than are non-polar compounds.

Explanations of the *shapes* of tlc spots can be given by consideration of adsorption isotherms (a topic introduced in *ACOL: Chromatographic Separations*).

Symmetrical spots are obtained if the isotherm is linear (Fig. 4.1e).

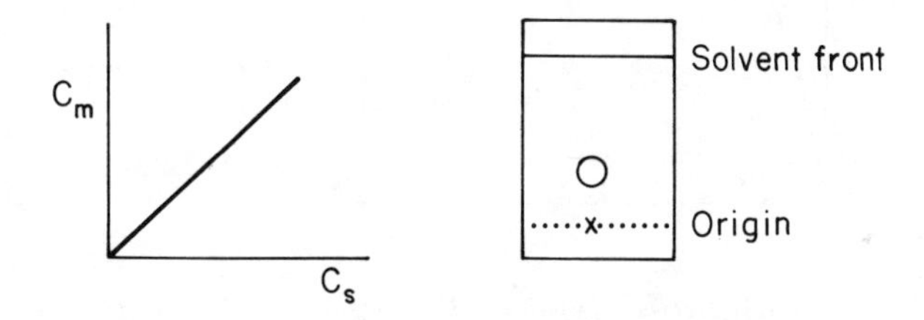

Fig. 4.1e. *A linear isotherm with corresponding chromatogram*

An S-shaped isotherm (Fig. 4.1f) indicates that after some molecules are adsorped on the surface, it becomes easier to adsorb others. So we find that molecules already adsorbed on the surface at the most active sites assist further adsorption by intermolecular bonding. Phenol forms a hydrogen bond first to the hydroxyl group of alumina and then the aromatic ring can help other molecules to associate by London Dispersion forces.

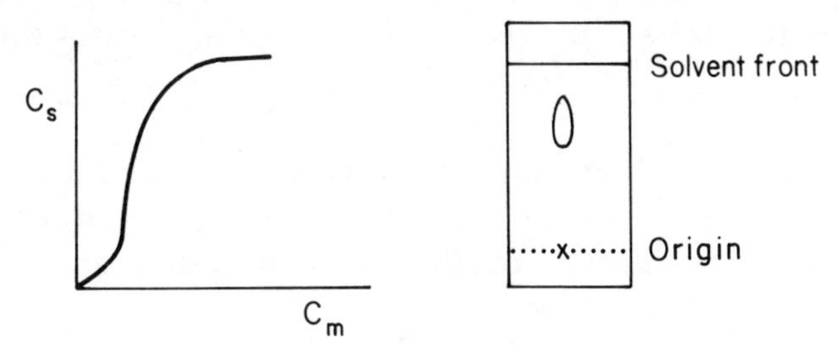

Fig. 4.1f. *An S-shaped isotherm with corresponding chromatogram*

This type of isotherm is associated with spots shaped like teardrops. A common spot shape is that exhibiting a tail. This arises when the isotherm is of the Langmuir type (Fig. 4.1g) because here the most active sites are first covered with solute molecules and the ease with which adsorption takes place decreases until finally the monolayer is complete when all the adsorption sites are occupied. The surface is said to be saturated.

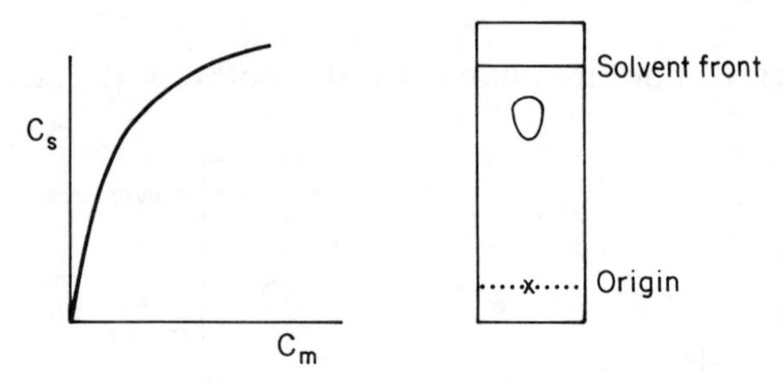

Fig. 4.1g. *Langmuir type isotherm with corresponding chromatogram*

A Langmuir isotherm is involved when there is no intermolecular bonding.

You might think from these considerations that it would be difficult to get symmetrical spots. However, in practice, we are using such small samples most of the time, that we are operating at the bottom of the isotherm curves where they are nearly linear. Distorted spot shapes are therefore often caused by overloading the plate and improvement can be achieved by reducing the sizes of the samples applied.

To help you decide how you will proceed when you are faced with an unknown mixture to separate, we have listed a number of typical mixtures in Fig. 4.1h, together with suggested combinations of mobile phase and adsorbent.

Glass	Adsorbent	Mobile Phase
aflatoxin	silica gel	methylbenzene : Ethyl ethanoate : propanone 3 : 2 : 1 + 1% HAc
aliphatic hydrocarbons	silica gel +20% $AgNO_3$	hexane (twice developed)
alkaloids	alumina	trichloromethane
amino acids	silica gel	propan-1-ol : 34% aqueous NH_3 67 : 33
aromatic hydrocarbons	acetylated cellulose	propan-1-ol : propanone : water 2 : 1 : 1
barbiturates	silica gel GF	ethyl ethanoate : methanol : aqueous NH_3 82 : 14 : 4
carboxylic acids	silica gel G	hexane : ethoxyethane : HAc 80 : 20 : 1
carotene	silica gel G	propanone : light petroleum 10 : 90
gangliosides	silica gel HPTLC	trichloromethane : methanol : water 2 : 1 : 1
metal ions	cellulose polygram cell 400	butan-1-ol saturated with 3*M*-HCl
monoterpenes	silica gel	benzene
nucleotides	cellulose	sat.aq. $(NH_4)_2SO_4$: 1*M*-NaAc : propan-2-ol 80 : 18 : 2
organochlorine pesticides	silica gel	heptane : propanone 98 : 2
organophosphorus pesticides	silica gel HPTLC	hexane : propanone 5 : 1
serotonin	silica gel	ethyl ethanoate : trichloromethane 3 : 2
sterols	alumina + $AgNO_3$	hexane : ethyl ethanoate 20 : 1
sugars	silica gel	propan-1-ol : 34% aqueous NH_3 67 : 33
sulphanilamides	silica gel	2-dimension 1. trichloromethane : methanol 95 : 1 2. ethyl ethanoate : methanol : aqueous NH_3 30 : 10 : 1
triacylglycerides	silica gel G	hexane : ethoxyethane : HAc 80 : 20 : 1
triacylglycerides	paraffin oil on kieselguhr	propanone : ethoxyethane 50 : 50 saturated with paraffin
vitamin K	silica gel	benzene
wax esters	silica gel	hexane : ethoxyethane : HAc 80 : 20 : 1

Fig. 4.1h. *Some typical mixtures, absorbents and suggested mobile phases*

Summary

Sample components can be placed in an order which relates to the way in which they interact with the adsorbent. Thus non-polar compounds interact by van der Waal's forces whilst more polar compounds interact by means of inductive and/or hydrogen bonding forces. The spot shape of a component can be related to the adsorption isotherm.

Objectives

You should be able to:

- explain the theoretical background to solute : adsorbent interactions;

- assess the interplay of R_f value and the type of solute : adsorbent interaction;

- recognise the shape of the adsorption isotherm which corresponds to symmetrical spots, teardrop-shaped spots and tailing spots.

5. Practical Techniques

5.1. APPLICATION OF SAMPLES TO THE PLATE (SPOTTING)

Liquid Samples – can be applied directly to the plate, but 1 to 10 μl of neat liquids will be too concentrated and will overload the plate. Liquids are generally mixed with a suitable diluent.

Solid Samples – cannot be applied directly to the plate so they must be dissolved in a suitable solvent. *The Optimum Concentration Range of These Solutions is 0.01 to 1.00 % w/v.*

Π *General Properties of Solvents*. The range of properties listed below applies to solvents used to dissolve samples and as mobile phases:

Cheap	Expensive
Analar Grade	Reagent Grade
Low Boiling	High Boiling
Unreactive	Reactive.

Underline those properties that you think may be important for the solvent used to dissolve the sample before spotting on to the plate.

There should be some overlap in the properties, eg it is very important to use *pure* solvents as any impurities in the sample solvent will contaminate the sample. The sample solvent should be *unreactive* towards the adsorbent and the solutes but cost is less important than in the case of a mobile phase. The more *volatile* the solvent the better, as each time you place a spot of solution on the plate, the solvent should evaporate quickly and evenly to produce a uniform spot. Adding more solution to a spot without allowing the solvent to evaporate causes an increase in spot size. It is particularly important to dry off the solvent before starting the development of the plate.

Choice of Solvents

Bearing in mind the above properties how do you choose the best solvent in which to dissolve the sample?

When a spot of solution is placed on an adsorbent surface, the solvent molecules compete with the sample molecules for active sites on the surface. If the solvent molecules are strongly attracted to the active sites the sample molecules move outwards to the edges of the spot and the result shown below is obtained on drying.

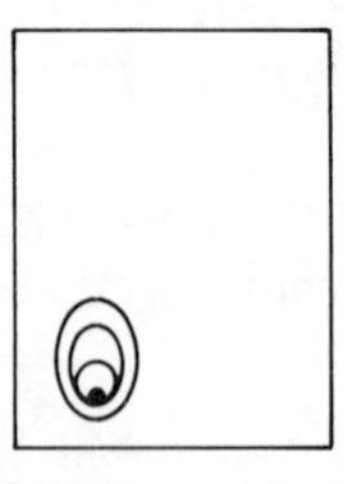

Fig. 5.1a. *The effect of spotting with a polar solvent*

SAQ 5.1a

How can we select a solvent with confidence from those in which the samples dissolve and be sure that we minimise interaction of the selected solvent with the adsorbent? Choose from the following ideas: ⟶

SAQ 5.1a (cont.)

(*i*) Select a solvent of minimum boiling point,

(*ii*) Select a solvent high in the eluotropic series,

(*iii*) Select a less reactive solvent,

(*iv*) Select a solvent low in the eluotropic series.

SAQ 5.1b

Given a choice of four solvents for a sample from the following:

cyclohexane, pyridine, chlorobenzene, water,

which would you choose as your ideal spotting solvent?

SAQ 5.1b

Amount of Sample Used

The sample solution should be made so that, in general, for an adsorbent thickness of 0.25 mm if you spot 10 μl of the solution on to the plate, the total sample loading will be from 1 to 50 μg. With a thicker adsorbent layer you can use more sample and in some cases, where you have a very sensitive method for detecting the solute you can use less.

Because of the difficulty of weighing very small quantities it is better to start with a solution of at least 2 mg per cm^{-3}. By making one or more dilutions it is possible to reduce the concentration of your solution to the range 0.1 to 500 μg per cm^{-3}.

SAQ 5.1c

Given a 10 cm^3 solution containing 20.0 g of a sample and diluting your solution by one hundred each time, how many dilutions are required to achieve a concentration of 0.2 g dm^{-3}?

Spotting the Sample on to the Plate

This task is easier where a pre-adsorbent layer is included on the plate (see Section 2.4) but the description below applies to plates without a pre-adsorbent layer.

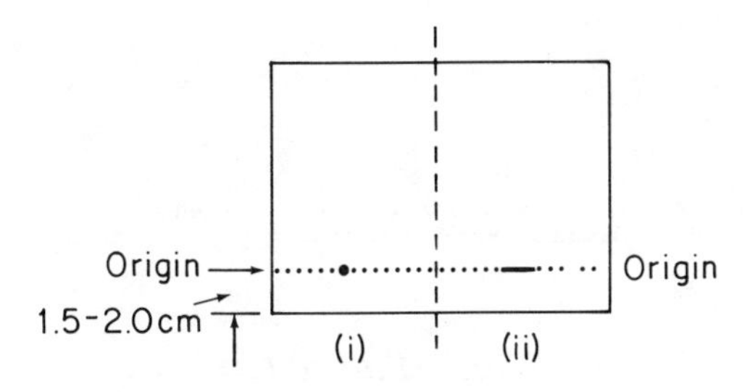

Fig. 5.1b. *Spotting a sample on to a tlc plate*

The sample should be spotted in one place and should have the smallest diameter possible (Fig. 5.1b (*i*)).

Π How can you ensure that spreading of the spot does not occur?

The best way to ensure that there is minimum spreading of the spot is to dry effectively between applications to the plate so as to remove the solvent.

Alternatively, the sample can be loaded on to the plate as a band (Fig. 5.1b (*ii*)).

Applications can be made with a capillary micropipette, which is easy to fill and may be calibrated to deliver precisely 1, 2, 5 or 10 μ and which is disposable afterwards. A syringe capable of delivering a known volume from 0.1 to 50 μ can also be used.

A plastic spotting template can help to ensure that the spots are evenly spaced across the plate. They come in various designs; the example illustrated below covers the plate, the sample being injected through holes cut at appropriate positions.

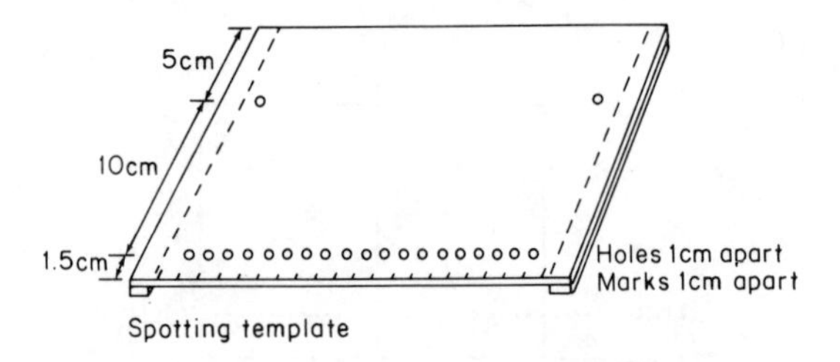

Fig. 5.1c. *Plastic template*

Precautions

It is important not to scratch the layer when the solute is applied as marks cause the mobile phase to elute unevenly and the spot can become distorted.

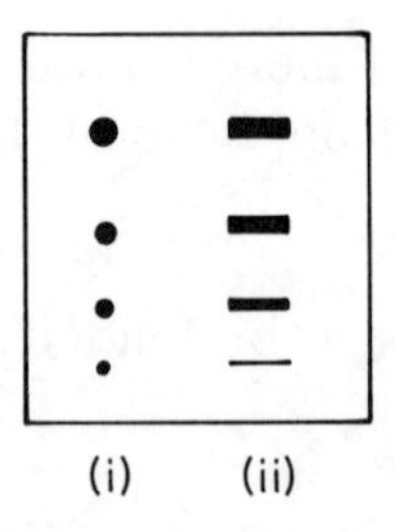

Fig. 5.1d. *A tlc plate where a satisfactory spotting technique was used*

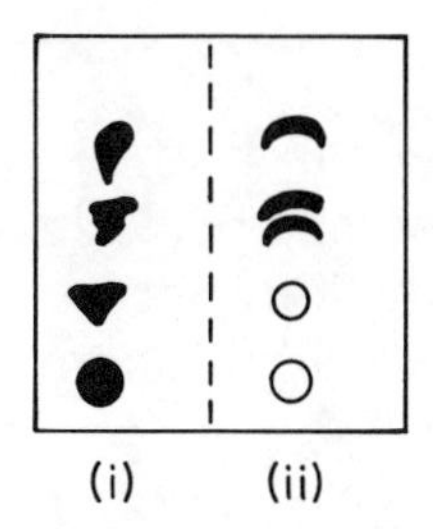

Fig. 5.1e. *A tlc plate where the spotting technique was unsatisfactory*

In Fig. 5.1d (*i*) and (*ii*) are examples of well spotted plates after elution while (*i*) and (*ii*) in Fig. 5.1e show:

(*i*) the effect using too much sample, ie too concentrated a solution. This effect is called tailing,

(*ii*) the effect of applying the sample without drying off the solvent between applications, so rings of sample appear.

Resolution and Spot Size

In a tlc system, resolution is calculated using the equation:

$$R_S = \frac{X}{0.5\,(d_1 + d_2)}$$

where X is the distance between the centres of two spots and d_1 and d_2 are the average diameters of the spots.

Components are *just* separated when $R_S = 1$.

So resolution can be improved by either increasing the value of X or decreasing the average diameter of the spots. It is the second factor that is important here and resolution can be improved if less material is put on to the plate. The sensitivity of the detection method is then the critical factor in determining the extent to which this is possible.

SAQ 5.1d

Resolution can also be improved by increasing X, the distance between spot centres. By reference to previous Sections where you think it is appropriate, is this improved resolution influenced by:

(*i*) the choice of adsorbent,

(*ii*) the shape of the chromatographic tank,

(*iii*) the nature of the mobile phase,

(*iv*) the nature of the spotting solvent?

SAQ 5.1e

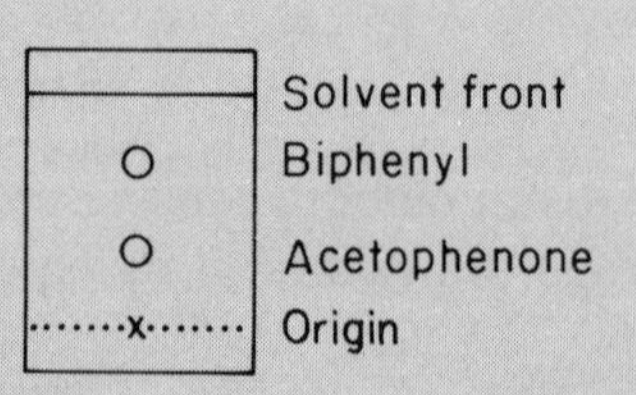

Fig. 5.1f. *Separation of biphenyl and acetophenone*

From the above diagram for the tlc separation of acetophenone and biphenyl, ⟶

SAQ 5.1e (cont.)

(*i*) Show that the resolution is equal to 2.5.

(*ii*) Choose from the following responses to explain how resolution could be increased.

(*a*) We could use a thinner layer on the tlc plate with the same amount of sample.

(*b*) The sample could be applied to the plate in a solvent such as methanol without allowing the methanol to dry between each application.

(*c*) We could choose a mobile phase with the same ϵ^0 value but with a different composition.

(*d*) We could put less sample on the tlc plate.

(*e*) We could choose a mobile phase with the same ϵ^0 value but with a different composition and also put less sample on the plate.

5.2. DEVELOPMENT OF THE CHROMATOGRAM

A prepared tlc plate coated with adsorbent, is activated and loaded as described in previous Sections. The plate must now be brought into contact with the mobile phase in a suitable tank (or a cylindrical jar for narrow plates).

A tank just big enough to contain the prepared plate is chosen, cleaned and lined with filter paper possibly inserted in a U-shape to line the two long walls. Mobile phase is added to a depth of 0.5 to 1.0 cm and, in order to saturate the air with respect to the vapour of the mobile phase, the tank is shaken to wet the filter paper then left at least 15 minutes to allow the atmosphere to reach equilibrium.

Π A worker feels that he could speed up the saturation process by adding a depth of 5.0 cm of mobile phase. However, his chromatograms do not develop and he has to look again at his technique to find the reason. Can you see his problem?

A depth of 5.0 cm of mobile phase will cover the samples spotted on the origin and they will simply diffuse into the solvent in the tank, so it is essential that the level of the solvent lies below the origin.

Sandwich Plates

In order to minimise the enclosed volume and encourage speedy saturation of the air over the plate, a sandwich tank or chamber consisting of a second glass plate on top of the loaded tlc plate can be used. The plates are held firmly by a clamp or rubber bands, care being taken not to disturb the adsorbent layer. A plastic or metal spacer may be used to keep the two glass plates separate.

Most development is done by upwards or ascending chromatography, ie the solvent is allowed to flow over the plate by capillary action. Since the solvent has to move against gravity, the rate of movement of the solvent front becomes slower for large plates.

Π Can you think of a situation in which the solvent would not have to move against gravity?

This can be achieved by placing the plate in a horizontal plane. The sandwich chamber can be used in this way, with the solvent being transferred from a trough to the adsorbent by means of a wick, usually filter paper, as shown in Fig. 5.2a.

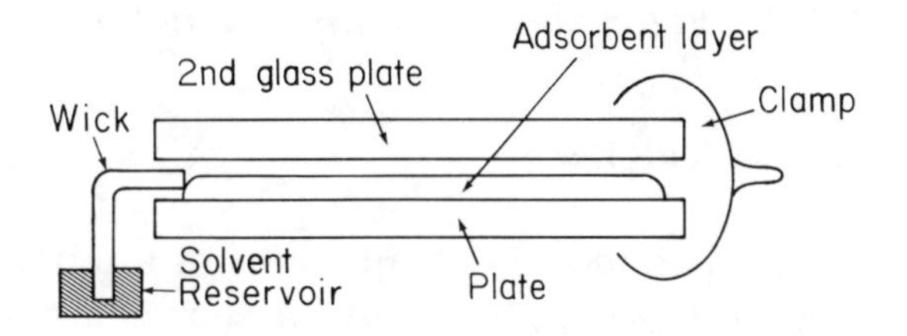

Fig. 5.2a. *Use of sandwich plate in horizontal plane*

Larger plates (40 × 20 cm) can be used in this way with components which are difficult to separate.

Two-dimensional Development

This method has been borrowed from paper chromatography and has been especially important for amino acids and carbohydrates. The sample is spotted at one corner of the plate 1 cm in from both edges. The plate is developed in the normal way for about 15 cm in one mobile phase. The result might be as shown in Fig. 5.2b. The plate is removed from the tank and dried. If the plate were now visualised it might look like the plate in Fig. 5.2b(*i*). However, the plate is not visualised at this stage. It is now turned through 90°and placed in a second tank containing a second mobile phase.

The components will have been partially separated by the first run and the line of spots which have been formed now acts as the origin for the second run (Fig. 5.2b(*ii*)).

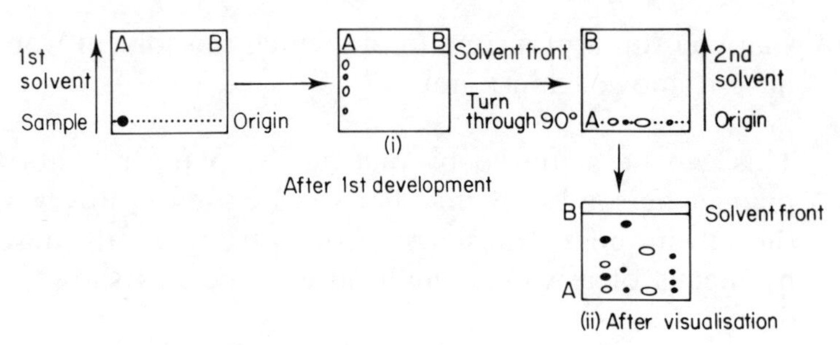

Fig. 5.2b. *Two-dimensional tlc*

5.3. VISUALISATION

Once we have separated the components of a mixture we need to be able to recognise where the spots for each component lie on the plate. The early workers tried to separate coloured compounds.

Unfortunately, few samples which we try to separate today are coloured. But as a general rule of visualisation we follow the example of the early workers and try to keep the detection method as simple as possible.

We can classify the methods available for visualisation in two ways. These are represented by two pie charts (Fig. 5.3a). This figure reminds us that the sub-divisions are not mutually exclusive, eg water can be in both the destructive and non-destructive portions.

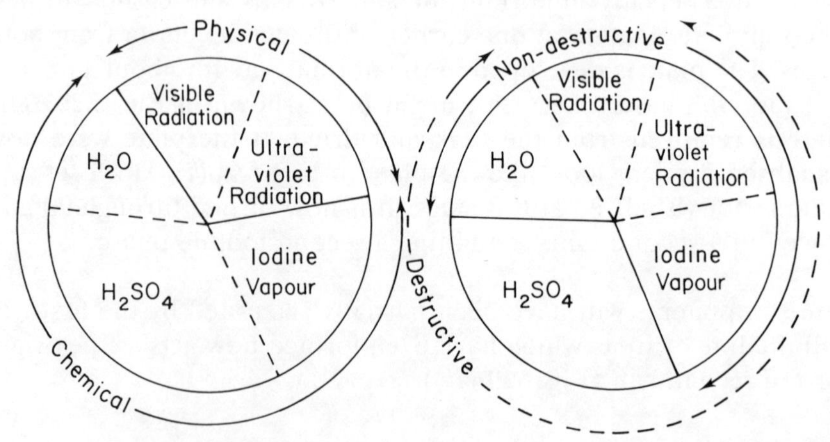

Fig. 5.3a. *Methods of visualisation*

Non-destructive Methods

Visible Radiation

When we look at coloured spots under visible light, we are using a non-destructive technique, ie if we wanted to remove the compound from the silica gel afterwards it would be unchanged by the action of the visualising technique. Examples of coloured substances are plant pigments, dyestuffs and food colourants

Ultraviolet Radiation

Many coloured compounds absorb uv radiation so we can locate the tlc spot by shining a uv lamp onto the plate. The spot usually shows up as a fluorescent area against a white background provided the silica gel contains no inorganic phosphor (eg zinc silicate or sulphide). If the phosphor is present, the spot shows up as a dark spot on a bright background.

Π Why has ultraviolet radiation also been included in the destructive part of the pie chart (Fig. 5.3a)?

This is because some compounds, such as certain vitamins, may undergo photochemical changes or decomposition on exposure to uv radiation. These examples are very much the exception and exposure to ultraviolet radiation is the most important non-destructive technique available.

We shall see in preparative tlc that we can spray the plate after separation with a dyestuff such as 2,7-dichlorofluorescein which facilitates visualisation under a uv lamp.

Iodine Vapour

To visualise with iodine is one of the simplest and most convenient methods. A tlc tank is set up with no solvent but with crystals of iodine in the bottom. Gentle heating will cause the iodine to vaporise and fill the air of the tank. When the tlc plate is brought into contact with this vapour, the iodine dissolves in the solutes which are revealed as brown spots of varying intensity.

Π On removing the plate from the iodine bath the outline of the spots should be marked by scoring the shape on the plate. From your knowledge of the properties of iodine can you suggest why?

The iodine will readily volatilise out of the spot and the plate will return to its all-white original condition within about half an hour.

A few substances, notably polyunsaturated fatty acids, *may* react chemically with the iodine so that the product after visualisation is not the same as the original fatty acid. For this reason it is not wise to use iodine to visualise a preparative separation where you do need to be sure that your product after visualisation remains unchanged.

Water

With many lipid or steroid substances, spraying with water can be an appropriate non-destructive visualising reagent. The sprayed plate is held against the light when the lipophilic compounds appear as white spots with a translucent background.

On occasions with these compounds, it is possible to see the spot as the plate is removed from the tank just before all the solvent has volatilised from the surface.

The pie chart (Fig. 5.3a) shows that water may be a destructive reagent because certain esters may be hydrolysed by spraying them with water. These compounds are exceptions rather than the rule.

Destructive Methods

Fig. 5.3b lists some of destructive reagents that react with the sample components on the plate to give coloured products which can then be seen in the visible region.

Fig. 5.3b. *Destructive visualising agents*

Samples	Reagent	Procedure	Result
Alcohols	Ceric Ammonium Nitrate	Dissolve reagent (6 g) in 4*M* HNO_3 (100 cm^3) Dry plate 5 min. at 105 °C. Cool before spraying.	Poly alcohols show brown spots on yellow ground
Alcohols (Bile Acids and Steroids)	Vanillin: Sulphuric Acid	Dissolve vanillin (3 g) in ethanol (100 cm^3). Add conc. sulphuric acid (0.5 cm^3) with stirring. Spray and heat at 120 °C.	Higher alcohols and ketones give blue–green spots
Aldehydes and Ketones	2,4-dinitro-phenylhydrazine	Dissolve reagent (0.4 g) in 2*M* HCl (100 cm^3).	Yellow/red spots
Alkaloids	Cobalt (II) thiocyanate	Dissolve ammonium thiocyanate (3 g) and cobalt (II) chloride (1 g) in water (20 cm^3).	Blue spots on white/pink ground

Fig. 5.3b cont.

Samples	Reagent	Procedure	Result
Alkaloids (antihistamines, cyclohexylamine, lactams)	Dragendorff (Munier modif) modification	Dissolve bismuth subnitrate (1.7 g) and tartaric acid (20 g) in water (80 cm^3) – Soln. (a). Dissolve potassium iodide (16 g) in water (40 cm^3) – Soln. (b). Spray reagent is prepared by mixing soln. (a) and soln. (b) (equal volumes). Take 5 ml of this soln and a soln. of tartaric acid (10 g) in water (50 cm^3).	Various colours
Amino Acids (also Amines)	Ninhydrin	Dissolve ninhydrin (0.20 g) in butan-1-ol (100 cm^3) – Soln. (a) 10% aqueous acetic acid – Soln. (b). Spray a mixture of soln. (a): Soln. (b) (95 : 5). Heat at 100 °C.	Pink-red spots on white background
Amines	Alizarin	Dissolve alizarin (0.10 g) in ethanol (100 cm^3)	Aliphatic amine and amino alcohols give violet spots on faint yellow background.
Barbiturates	*s*-Diphenylcarbazone	Dissolve *s*-diphenylcarbazine (0.10 g) in 95% ethanol (100 cm^3)	Purple spots

Carbohydrates	*p*-anisaldehyde	Dissolve *p*-anisaldehyde (1 cm^3) and conc. H_2SO_4 (1 cm^3) in ethanol (18 cm^3). Spray and heat at 110 °C.	Sugar phenylhydrazone give green–yellow spots
Carboxylic Acids	*p*-anisidine-phthalic acid	Dissolve *p*-anisidine (1.23 g) and phthalic acid (1.66 g) in methanol (100 cm^3)	Aldohexoses green, pentoses red–violet, methylpentoses yellow/green, uronic acid brown
	Bromocresol green	Dissolve reagent (0.04 g) in ethanol (100 cm^3). Add 0.1*M* NaOH until blue colour just appears.	Yellow spots on green background
Esters and Amides	Hydroxylamine/Ferric nitrate	Dissolve hydroxylamine hydrochloride (1 g) in water (9 cm^3) – Soln. (a). Dissolve sodium hydroxide (2 g) in water (8 cm^3) – Soln. (b). Dissolve ferric nitrate (4 g) in water (60 cm^3) and acetic acid (40 cm^3) – Soln. (c). Spray with a mixture of one volume soln. (a) and one volume of soln. (b). Dry at 110 °C for 10 min. Spray with a mixture of soln. (c) (45 cm^3 and conc. HCl (6 cm^3).	Coloured spots

Fig. 5.3b cont.

Samples	Reagent	Procedure	Result
Lipids	Rhodamine B	Dissolve rhodamine B (0.05 g) in ethanol – Soln (a) 3% hydrogen peroxide – Soln. (b) 10*M* KOH – Soln. (c). Spray with Soln. (a). Observe in visible and uv radiation. Spray with Soln. (b) to enhance colour.	Bright-red fluorescence
Lipids	Rhodamine B	Spray with Soln. (a) and then Soln. (c)	Triglycerides yield bright white spots on a pink-red background
Phospholipids	Dragendorff's reagent	Dissolve 17% basic bismuth nitrate in 20% aqueous acetic acid. Soln. (a) 40% aqueous potassium iodide Soln. (b) water Soln. (c). Spray Soln. (a):(b):(c):4 : 1 : 14.	Phospholipids containing choline give orange colours.
Pesticides	Diphenylamine: zinc	Dissolve diphenylamine (0.5 g) and zinc chloride (0.5 g) in acetone (100 cm^3). Spray and heat at 200 °C for 5 min.	Chlorinated pesticides give various colours

	Brilliant Green	Dissolve brilliant green (0.5 g) in propanone (100 cm^3). Spray and expose plate to bromine vapour.	Organophosphorus and triazines give green spots
Phenols	Ammonium vanadate: anisidine	Saturate water with ammonium vanadate – Soln. (a). Dissolve *p*-anisidine (0.5 g) in conc. H_3PO_4 (2 cm^3), dilute to 100 cm^3 with ethanol and filter – Soln. (b). Spray with Soln. (a), while plate is wet, spray with (b). Heat at 80 °C.	Various coloured spots on a pink background.
Steroids	Anisaldehyde: antimony trichloride	Mix *p*-anisaldehyde (1 cm^3) with 100 cm^3 of saturated antimony trichloride in chloroform. Add conc. H_2SO_4 (2 cm^3). Keep soln. at room temperature in dark for 1.5 hours. Remove upper layer of reagent mixture and use it to spray plate. Dry plate 5 minutes in dark and heat at 90 °C for 3 minutes. Observe in visible light and uv.	Various colours

Most organic compounds will char when sprayed with 50% concentrated sulphuric acid and then heated to 110 °C. On the tlc plate brown or black spots are produced which are often used when quantitation is required.

An alternative reagent is a 5 per cent solution of potassium dichromate in 40% sulphuric acid. This reagent can be sprayed onto plates which are then heated to 110 °C for up to 15 minutes.

SAQ 5.3a

Indicate whether each of the following statements is TRUE or FALSE.

(*i*) In visualising an amino acid separation by spraying the plate with ninhydrin, a research worker was using a destructive technique.

TRUE / FALSE?

(*ii*) After separating fatty acids on a silica gel GF, a research worker examined the spots under an ultra violet lamp. A chemical method of visualisation was being used.

TRUE / FALSE?

(*iii*) In preparative tlc a worker says destructive visualisation techniques are preferable to non-destructive techniques.

TRUE / FALSE?

(*iv*) You can use water as an appropriate non-destructive visualising agent with esters.

TRUE / FALSE?

(*v*) Inorganic compounds can be visualised by using sulphuric acid as a destructive agent.

TRUE / FALSE?

⟶

SAQ 5.3a (cont.)

(*vi*) Ultraviolet radiation can be used as a non-destructive visualisation technique only with coloured substances.

TRUE / FALSE?

Summary

The properties needed for a spotting solvent are not identical to the properties needed for an eluting solvent though there is some overlap. The sample can be added to the plate as a spot or as a streak. The separation can be achieved in normal chambers or in sandwich tanks. An improved separation may be seen when two-dimensional chromatography is performed. The separated spots may be visualised by a physical or a chemical agent. An alternative subdivision of visualising agents is 'destructive' and 'non-destructive' modes.

Objectives

You should be able to:

- recognise the cause of mishapen spots;
- calculate the dilution needed to obtain a solution which gives the required quantity of sample on the tlc plate;
- set up a chromatographic system for descending tlc;
- use Fig. 5.3a to determine which visualising agent you would use for a given compound.

6. Applications

6.1. REPRODUCIBILITY OF R_f VALUES

Π You have been given an unknown organic compound in a laboratory exercise. You can see it is a solid. What physical constant would help you to identify it?

A melting point is the constant that probably first comes to mind and generally the melting point and/or boiling point of the unknown and the melting point of a derivative, together with spectroscopic properties can all help in the identification.

So can tlc. If we have given the impression to date that tlc is useful only for separating the constituents of a mixture we will rectify this now and show how tlc can help to identify an unknown substance. Some workers claim that the R_f value should be added to the 'classical' properties quoted for organic compounds, and if, as some claim, R_f values accurate to ± 0.05 can be measured then it should be a powerful addition to the list.

Most workers, however, do not have such faith in measured R_f values and they think of them as a 'guide' which can be used along with reference standards and/or colour reactions produced by specific spray reactions. The lack of faith in R_f values stems from a recognition that when a known compound is examined by tlc on one day its R_f value may differ from this value if measured on another day.

Π A worker sets out to measure the R_f value of a single compound. He uses a manufacturer's pre-coated plate following instructions with respect to activating it. He makes up a mobile phase and allows his tlc tank to saturate over lunchtime before inserting and running the chromatogram, keeping a constant eye on the proceedings. His measured R_f value is 0.52. Referring to a book, he sees a quoted value of 0.48 for the same substance. The next day he decides to check his value in the hope of getting nearer 0.48. He discovers he has run out of prepared plates so he makes his own, but in his general hurry he activates his plates for only 15 minutes and uses yesterday's mobile phase, giving his tank only 5 minutes to saturate in order to run the chromatogram over lunchtime. His R_f value is 0.45. He is happy to be nearer the book value.

Is his happiness justified and which R_f value is correct?

His happiness is clearly not justified as each value is 'correct' with respect to the conditions used and the book value needed to be carefully considered with respect to the conditions, plates, mobile phase and experimental details as described by these workers.

It is important to look at the following factors which our careless worker did not take sufficiently into account and changes in any one of these can alter the R_f value quite considerably.

(*a*) The solvent vapour (chamber saturation).

(*b*) Quality and quantity of mobile phase.

(*c*) Adsorbent activity.

(*d*) Chromatographic techniques and conditions.

Variations and lack of reproducibility are generally caused by not being sufficiently careful about the above conditions in separation.

(*a*) Solvent Vapour (Chamber Saturation)

To obtain reproducible R_f values you must ensure that the atmosphere in the tank is saturated with respect to the solvent vapour.

SAQ 6.1a

Which of the following would help to ensure a saturated atmosphere in a tank?

(*i*) Leaving the tank for 5 minutes or less before developing the plate (refer our worker on day two).

(*ii*) Leaving the tank for an hour or more before developing the plate (refer our worker on day one).

(*iii*) Lining the tlc tank with filter paper.

(*iv*) Using a larger tlc tank with respect to the size of the plate.

(*v*) Using as small a tank as possible with respect to the size of the plate.

If the chamber is not saturated, the solvent rising up the tlc plate evaporates from the surface to try to saturate the air and the higher up the plate the more evaporation takes place. This can lead to the very peculiar results shown in the Fig. 6.1.

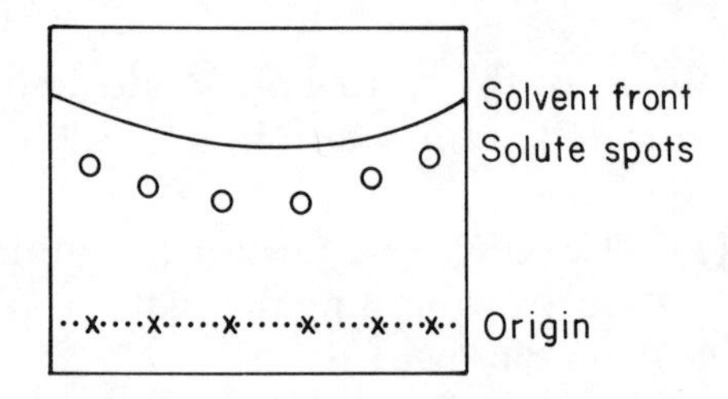

Fig. 6.1. *Tlc plate developed in an unsaturated tank*

In Fig. 6.1, one compound has been spotted at the origin in six places. The solvent vapour in the air is not sufficient to saturate it, so solvent evaporates from the plate doing so more rapidly from the edges (where there is more air to saturate) than from the centre as the solvent rises. A concave solvent front develops and the R_f value is quite different if read at the edge as compared to the value read at the centre of the plate.

With mixed solvents, saturation of the air with vapour becomes even more important and in extreme cases two solvent fronts may develop giving very variable R_f values.

It was to minimise these difficulties that Stahl developed the sandwich tanks as described earlier, where the inner air gap is only 1–2 mm, to facilitate rapid saturation with vapour.

(*b*) Quality and Quantity of Mobile Phase

Bearing in mind the qualities already listed as being of importance in the mobile phase, purity is the one most relevant to reproducibility.

SAQ 6.1b

What change in ϵ^0 value would you expect to find if dichloromethane contained 1% methanol, compared to pure dichloromethane?

Also, with solvents that are fairly volatile it is best to use freshly prepared solvent mixtures for each chromatographic run since volatilisation of one or more of the components will alter the overall solvent composition and hence the resulting R_f values. This factor may have affected our earlier worker's results as he used the same mixture of solvents over a two day period.

(*c*) Adsorbent Activity

Activity or *adsorption capacity* depends on the amount of water in the adsorbent layer.

Π How can you control the amount of water in a *home made* plate before use?

By heating or activating the plate, but note that consistency is all important – heating to 110 °C for 30 minutes for every plate is the norm.

Note: Overheating silica gel at 200 °C, where the activity depends on the silanol group (SiOH) causes loss of these groups and their conversion to siloxane groups (Si—O—Si). Resultant R_f values will be quite different.

However, activated plates on exposure to air can pick up moisture and within a few minutes lose almost all their activity – eg with a relative humidity of 50%, an active plate can lose 50% of its activity in three minutes. R_f values can vary by as much as 300% between runs done with a relative humidity of 1% compared to one of 80%.

To summarise, you have to heat the plate to activate it, but you can't handle it hot and you do need to spot samples on it, so it has to be cooled and exposed to the air which can cause deactivation and subsequently alteration to R_f values.

Π How can you get round the relative humidity problem?

Decide which of the following approaches would be most appropriate in your environment at work:

(*i*) All plates, once prepared can be dried in an oven and stored in a constant humidity cabinet at 50% relative humidity, then spotted and handled as quickly as possible before development.

(*ii*) The plates can be prepared in the normal way and spotted, then reheated to 100 °C for 30 minutes, but this is only possible where the samples are high boiling. This would not be appropriate for use with eg terpenes in the perfume industry.

(*iii*) The plates, once prepared and activated can be stored in a desiccator and all handling done as quickly as possible.

(*iv*) The plates can be handled in a special room with controlled temperature and humidity, if you are lucky enough to have access to such facilities.

(*v*) The use of some ready-coated plates may remove the need for activation as the manufacturer knows that on opening the pack, all the plates will become fully saturated with respect to water vapour and you will get reproducible R_f values.

The important feature is to standardise your own procedure.

Quality of Adsorbent

Not only the treatment of the plates, eg activation, can alter R_f values but an adsorbent, eg silica gel, can vary in quality from manufacturer to manufacturer. The main characteristics of the adsorbent that need to be compared are particle size, pore volume, pore diameter and surface area.

However, since R_f values can change from one product to another, it is probably better to stick to one type of plate and in some difficult separations you may find one specific manufacturer's plates reported as achieving satisfactory separations whilst five other manufacturer's plates did not.

If you do decide to change from one manufacturer's products to another you can always ask the sales representative to prove that their product can do the job before ordering a large supply of plates.

Layer Thickness

Theoretically, R_f values are independent of the layer thickness if you keep the other variables constant. Therefore taking pre-coated plates of 0.10 mm layer thickness and self-prepared plates of 0.25 mm thickness the R_f values should not vary. In practice, there is likely to be a difference in R_f values not caused by layer thickness but by pore size.

(*d*) Chromatographic Techniques and Conditions

Variables here which might affect R_f values are (*i*) methods of developing the chromatogram, (*ii*) temperature, (*iii*) distance over which chromatogram is run and (*iv*) amount of sample used.

(*i*) Methods of developing the chromatogram

Π All other conditions being kept constant do you think R_f values will depend on whether the chromatogram is developed using ascending, descending or horizontal methods?

In practice, all seem to give similar values.

(*ii*) Temperature

Π Is the temperature at which you run the chromatogram important and, if it varies, does it affect R_f values?

This is not easy to answer, there being much controversy over whether R_f values are affected by temperature. Some workers claim that temperature fluctuations cause significant variations in R_f values whilst others claim the values are unaffected. As a rule, it is better to operate at constant temperature and for very accurate work it may be necessary to insulate the tank so as to ensure that development occurs with as little variation as possible.

(*iii*) Distance moved by solvent front

With a single solvent system, R_f values are independent of how far the solvent front moves.

SAQ 6.1c

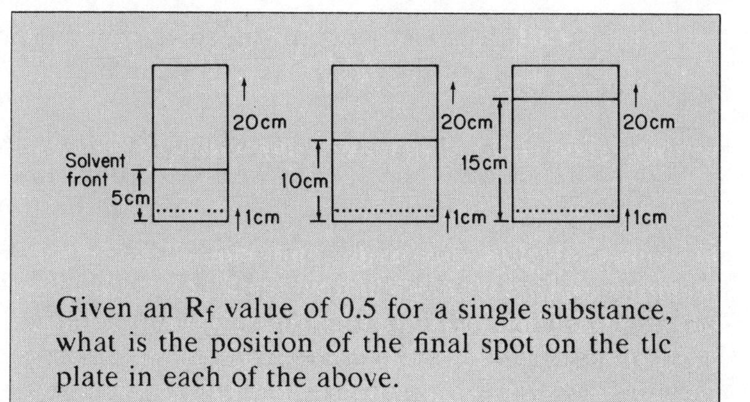

Given an R_f value of 0.5 for a single substance, what is the position of the final spot on the tlc plate in each of the above.

In practice it would be important to take note of the following points:

(*a*) If a tank is not properly saturated, a longer running time produces more distortion of R_f values.

(*b*) Assuming saturation of the atmosphere in the tank is complete, the distance between the origin and the solvent finishing line (solvent front) should be kept constant to ensure maximum reproducibility.

(*c*) The distance between the solvent immersion line and the origin should also be kept constant.

(*d*) Finally, one bad habit which has often been recommended is the scoring of a line 10 or 15 cm from the origin. The idea was to have a position from which to measure R_f values. Unfortunately, it is easy to forget about the tlc plate and go off to lunch. On return the solvent front has been at the scored line for an indeterminate time.

(*iv*) Sample loading

The effect of sample size varies depending on whether the components have convex or concave isotherms.

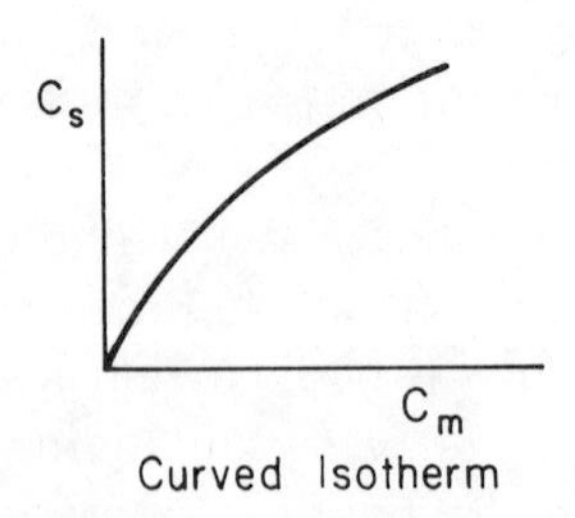

Curved Isotherm

Π Is the above a concave or a convex isotherm?

It is a convex isotherm and when a solute has a convex isotherm, R_f values increase with increasing sample load. Remember that the shape of the spots produced with too much sample is elongated with a tail:

When the solute has a concave isotherm , R_f values decrease with increasing loads.

6.2. QUANTITATIVE tlc

Tlc, so far, has been represented as a cheap, easy to use, *qualitative* technique, and it can be all of these, but it can also be made *quantitative* thus introducing a new dimension. Not only can you determine what constituents are in a mixture, but you can also measure how much of each constituent is present.

Quantitative determinations are based on one of the following principles:

1. Separation of the sample on the tlc plate followed by elution of the individual constituents and subsequent quantitation.

2. Quantitation *in situ*, on the plate.

Quantitative tlc involves great care in ensuring reproducible conditions. It is essential that the spot moves to the same part of the plate in each run to ensure uniformity of size of the spot and especially if spot area is being used as the basis of quantitation.

Π Assuming that reproducible R_f values are important, which of the following factors would influence the results?

1. Time of activation of the plate.

2. Storage of plate in desiccator.

3. Saturation of the atmosphere in the tank.

4. Temperature used for development.

Referring to Section 6.1, all four factors are of vital importance in ensuring a constant and reproducible R_f values.

R_f Values in Quantitation

For best results in any of the quantitative methods, the spots should have an R_f value between 0.3 and 0.7. Spots with low R_f values, ie below 0.3, are too concentrated whereas those with R_f values above 0.7 are too diffuse. At either extreme, estimation is unreliable so you cannot make comparisons or estimates of the concentrations of four components whose R_f values are 0.1, 0.3, 0.5 and 0.8. You can quantitate across the plate horizontally but not generally up and down the plate vertically.

Quantitative Analysis Based on Principle 1

Having developed a technique to ensure reproducible R_f values, the next stage is to remove the solute from the adsorbent and assess the amount present. Generally, if you need to determine the amount of solute in any spot you need to visualise it first.

Π Given the choice of a destructive visualising agent such as conc. H_2SO_4 or a non-destructive one such as ultraviolet radiation, which would you choose?

A non-destructive visualisation method is essential otherwise you affect the material in the spot and you cannot then determine how much there was.

After visualisation one of the following procedures could be used:

1(a). Visualisation by a non-destructive technique

↓

Scrape off spot from glass plate.

↓

Transfer to centrifuge tube.

↓

Add suitable solvent to dissolve solute.

↓

Stir to dissolve solute.

↓

Spin tube in centrifuge.

↓

Remove supernatant liquid.

↓

Collect supernatant liquid and apply a suitable quantitation technique (see below).

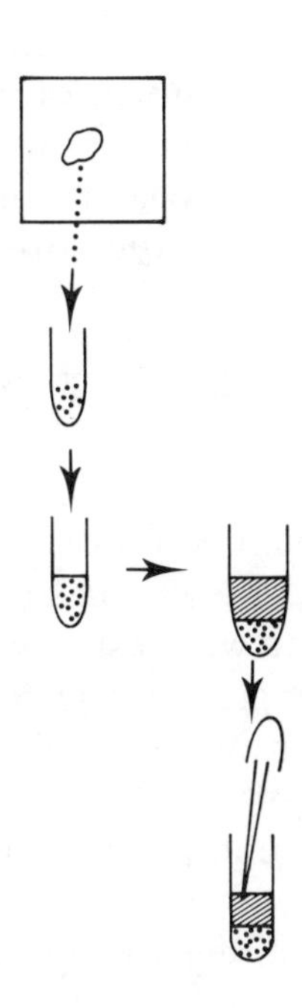

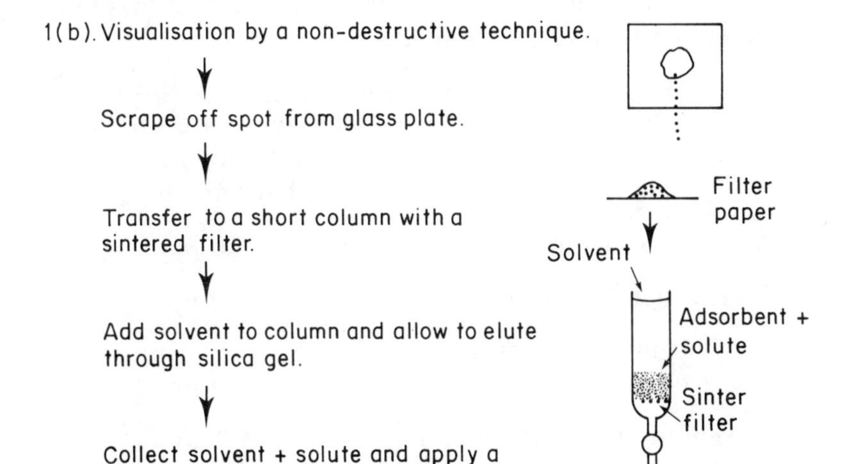

Π Do you think the choice of eluting solvent could affect the amount of solute recovered?

If you think *yes* is the correct answer, you're right. The choice of eluting solvent is particularly important. For example, the steroid cortisol can be eluted from silica gel using benzene and water but only 1% is recovered, while the use of dichloromethane : methanol 9 : 1 gives 90% recovery.

Quantitation Techniques

The simplest technique to apply to the resulting solution from either method 1(*a*) or 1(*b*) is to separate the solvent by distillation and weigh the extracted solute. This can be applied where a full plate with a streak of 20 mg has been applied. One advantage is that this gravimetric technique does not require any calibration. However, there are disadvantages in that most silica gels contain organic impurities which can be extracted with most organic solvents and these impurities augment the mass determined.

By either of the two methods 1(*a*) and 1(*b*), you finish up with a solution of the solute of unknown concentration. So the next task is to determine this concentration. You will need to obtain calibration curves for known quantities of the solute in the chosen solvent(s) and use them to analyse the samples.

Technique	Mass of Solute which can readily be determined
Visible Spectrometry	100 μg
Ultraviolet Spectrometry	50 μg
Nuclear magnetic resonance Spectrometry	10,000 μg
Gas–liquid Chromatography	1 μg

Fig. 6.2a. *Level of sensitivity for quantitation*

Quantitative Analysis Based on Principle 2

Here the analysis is completed *in situ* so obviating need for an extraction procedure and the problems of choosing solvents and removing spots.

This can be done in the following ways.

(*a*) Visual Inspection

A series of standard solutions with known concentrations of the solute is spotted onto the plate, alongside sample spots, the same volumes of sample and standard being used each time. The spot area should be kept as small as possible.

∏ After making up a standard containing 5 μg, an analyst decides to spot twice as much solution to get a 10 μg standard, 4 times as much for a 20 μg and 8 times as much for a 40 μg standard. Would this be acceptable?

No. This procedure would not be acceptable. The same volume of solution of sample and standard needs to be used. So the analyst should have made the 40 μg concentration and diluted this to obtain the others, then used the same volume of each solution.

∏ How can the spot area be minimized?

A minimal spot area can be ensured by adding one drop of solution at a time and allowing the solvent to evaporate before applying a second drop.

Running standards and samples on the same plate ensures that the same solvent is flowing over all spots, and variations in layer thickness will be lessened by comparing spots with the same R_f values.

The results obtained by visual inspection can be accurate to within $\pm$ 10% if great care is taken and skill executed, and for many analyses in industry and research, this semi-quantitative technique can be accurate enough. It can also be used to decide on the presence or absence of an impurity. If a more accurate method is required, the measurement of spot areas can be investigated.

(*b*) Measurement of Spot Areas

This method, being widely used in paper chromatography, was the first to be tried in tlc and as with any quantitative measure it depends on:

1. the reproducibility of the silica gel layer,

2. the ability to apply the sample and standard in a reproducible fashion (compare with (*a*)),

3. a mobile phase that produces well separated spots with no tailing.

The square root of the area of the spot is proportional to the logarithm of the concentration of the solute in the sample solution. However, the diffuse boundaries of tlc spots reduces the accuracy of the measurements.

(*c*) Quantitative Densitometry

This is defined as resolving a compound on a tlc plate, visualising the spots and measuring the absorbance of each spot directly on the plate. The amounts of material in the unknowns are measured by comparing them to a standard curve from reference standards that are separated under the same conditions.

Basic densitometer design includes a uv and/or visible radiation source, focussing optics and a photomultiplier detector.

This instrumentation can be very expensive and an integrated function of spot area is measured. The difference in intensity between the incident and transmitted radiation is measured and fed to a chart recorder after electronic processing. Scanning across a series of tlc spots produces a chart record rather like a gas or liquid chromatogram.

The height of the peak is a measure of the intensity of the spot whilst the peak width is proportional to the spot length in the direction of scanning. Over a limited concentration range, a linear relation exists between peak area and square root of the mass of solute.

Using an Internal Standard

If a known compound is added to the unknown mixture at a fixed concentration, the added compound can act as an internal standard.

The analysis does not then depend on applying samples of equal size to the plate.

The ratio of sample (peak height)/internal standard (peak height) is used in preparing a calibration graph.

These ratio calibrations are constructed by chromatographing aliquots of mixtures containing the compound of interest in various concentrations together with a constant concentration of the internal standard.

The peak heights of the components of interest are then determined. Ratios of component (peak heights)/internal standard (peak height) are plotted against concentration.

This plot is linear and can be reproduced with a precision of 0.7 to 1.0%. The internal standard must be well separated from other zones on the tlc plate. Clearly the internal standard must not be present in the original sample.

6.3. PREPARATIVE tlc

In preparative tlc the emphasis is on, not only the analysis of a mixture of solutes, but the preparation and collection of samples of the individual constituents.

Π What do you think would be the most useful modification to tlc if the aim is the preparation and collection of individual samples?

Clearly it would involve the use of much larger samples and to achieve the separation of larger samples you need to modify:

(*a*) the adsorbent layer,

(*b*) the method of applying the sample to the plate,

(*c*) the visualisation technique,

In addition the separated constituents will need to be removed from the plate.

(*a*) *The Adsorbent Layer in Preparative tlc*

In order to apply larger samples to the plate, you would need to change the thickness of the adsorbent.

Π Would you increase or decrease the layer thickness?

For larger solute samples you would increase the layer thickness.

You will remember that analytical separations are performed on a 0.25 mm layer so for preparative purposes you will look for layers much thicker than this. However, you must exercise care that you are not carried away with enthusiasm for this modification – 'the thicker the layer, the more sample I can handle'.

Thick layers of 2.0 mm are fragile and the resolution is often poorer than that achieved on an analytical scale.

A reasonable compromise between increasing the amount of adsorbent and separation efficiency is found, by many workers, to be to use 0.50 mm thickness with a 20 mg sample. Trial and error will indicate the maximum possible sample loadings in a particular case for a required degree of separation. So whilst 100 mg may be suggested for a 1.00 mm silica gel layer (20 cm plate), it may be sensible to use less, if a good separation is required.

(*b*) *Application of the Sample*

A 10 μl syringe can be used to apply sample to the plate but repeated applications are necessary and time-consuming.

With very thick layers, troughs can be cut half as deep as the layer, and the sample placed in the groove, but this is also time-consuming.

The most usual way of applying the samples is by streaking. A larger 250 μl or 500 μl syringe is used and a continuous band of solute is streaked along the origin. Automatic streaking equipment which results in better band shapes for the separated components is available.

Pre-coated plates with a pre-adsorbent layer, where up to 5 cm^3 of sample can be applied without very careful spotting or streaking, can also be useful.

(*c*) *Visualisation Technique*

Π Remembering that preparative tlc seeks to facilitate the separation and recovery of individual compounds, would you use a destructive or a non-destructive technique for visualisation?

If there is a choice, a non-destructive technique will be best to prevent the destruction of the compound which you are interested in. But if the sample cannot be visualised non-destructively, all is not lost, although a little separated compound may be!

The plate can be sprayed with a visualising agent along the edges using a mask to protect the centre of the plate. The bands become evident at the edges, and we can assume that the band stretches across the plate so the central section can be scraped off with the unaffected compound.

Π Why do we not simply run one plate, measure the R_f values and use these R_f values for subsequent plates?

You will recall in Section 6.2 that we stressed how much care is required to ensure consistency in R_f values. Most workers are not prepared to go to such lengths so the method of spraying part of the plate is preferred.

Extraction of the Separated Bands

The simplest method is to scrape the outlined band of adsorbent + component from the plate. The adsorbent is then placed in a small chromatographic column having a sintered glass filter in the bottom and the component solute eluted from the silica gel using a suitable solvent. An alternative method involves placing the adsorbent in a centrifuge tube to which you add the solvent. The adsorbent is slurried in the solvent with a stirrer and when the extraction is complete the tube is placed in a centrifuge. After spinning, the tube will have the adsorbent as a cake in the bottom from which the solution of the component can be removed by Pasteur pipette, see p.85.

To minimise the dangers of the adsorbent blowing away after it has been scraped from the plate, small recovery tubes can be purchased which will operate with a vacuum pump.

Summary

To obtain the separations which give reproducible R_f values, it is necessary to standardise the adsorbent and its activity, the mobile phase, the chamber saturation and the mode of chromatographic development.The separated components of the sample are extracted from the adsorbent with a solvent. Subsequent quantitation can be done gravimetrically or spectrophotometrically. Alternatively the sample can be determined on the plate by visual inspection or by spectrophotometry or by measurement of spot areas. In preparative tlc, an increased layer-thickness is used and different methods of sample application may be required.

Objectives

You should be able to:

- choose a chromatographic system to permit the best separation with optimum reproducibility of R_f values.
- recognise the limitations of quantitative tlc;
- assess the appropriateness of quantitating the sample on the plate or after extraction from the plate;
- adapt a tlc system for preparative tlc.

7. Recent Developments

7.1. REVERSE PHASE tlc

One of the major advantages of tlc is that it is very easily modified. Early workers recognised that the silica gel could be used as a support material and a liquid phase could be coated onto it. This was achieved by placing the tlc plate in a tank containing a solution of 10% *n*-tetradecane in light petroleum and allowing this solution to flow over the plate as in normal development. When the solution has reached the top of the plate, the plate is removed and the solvent allowed to evaporate to leave the silica gel particles coated with *n*-tetradecane. The method of tlc we have previously described is called normal phase, ie we have a polar adsorbent as the stationary phase, eg silica gel or alumina and a less polar solvent as the mobile phase used for the elution, eg hexane. In reverse phase tlc the *n*-tetradecane on silica gel is the stationary phase which is very non-polar and now the mobile phase is the polar part of the system being propanone or acetonitrile which has been saturated with *n*-tetradecane.

Π Why do we need to modify the solvent by saturating it with *n*-tetradecane?

This is essential since otherwise the acetonitrile would remove the *n*-tetradecane from the silica gel and we would drastically alter the stationary phase.

This form of chromatography does not involve adsorption. Here we are dealing with partition chromatography, ie the sample components partition between the stationary phase and the mobile phase. We have already met an example of this earlier when discussing the use of cellulose as a stationary phase because water on the surface of the cellulose acts as the partitioning medium.

Separations depend on the balance between polar and non-polar groups in the mobile and the stationary phase as well as in the solutes.

It has been found that partition chromatography is better for separating mixtures according to the chain lengths of the components, the retention of homologues increasing exponentially with carbon number.

Although the sorption mechanism is assumed to be one of partition, much discussion is still being expended on the contribution of the silanols on the surface of the silica gel to the retention of the molecules. It seems likely that for octadecyl chains, the van der Waals attraction between the hydrocarbon part of the sample molecule and the octadecyl chain on the tlc layer is involved.

The method of preparing the tlc plate by treatment with *n*-tetradecane can still be used but it is much more usual to-day to purchase plates ready-made. The manufacturers followed the lead of researchers who recognised that it would be better to have the hydrocarbon chain chemically bound to the silica particles so as to stop the leaching of the hydrocarbon by the mobile phase.

The hydrocarbon chain is prepared by reacting silica with a suitable mono, di or tri-chlorosilane

eg
$$-Si-OH + Cl-\underset{CH_3}{\overset{CH_3}{\underset{|}{\overset{|}{Si}}}}-R \rightarrow -Si-O-\underset{CH_3}{\overset{CH_3}{\underset{|}{\overset{|}{Si}}}}-R + HCl$$

Any Si—OH groups that have not reacted with the chlorosilane and would act as adsorption sites are further reacted (capped) with hexamethyldisilazane.

Usually the R group is $CH_3(CH_2)_{17}$—, but octyl, $CH_3(CH_2)_7$—, and ethyl CH_3CH_2— have also been prepared. They are usually denoted RP-18, RP-8 and RP-2.

Diphenyl plates are also available. Each of these adsorbents has slightly different properties and each will need to be tested for the separation you may wish to achieve.

Sample Application

The sample is applied in the same way as samples in adsorption chromatography. Where possible we use a pre-adsorbent layer since this allows a fairly crude sample to be put on to the plate.

The plates do *not* need to be activated by heating prior to use, but it may be worth while washing them by allowing a mixture of $CHCl_3$: MeOH(1 : 1) to flow over the adsorbent without applying any sample. The layer is then dried before the samples are spotted onto the plate.

If possible, the samples should be dissolved in methanol because it evaporates readily. Avoid water since it tends to spread the spots over a wider area.

Mobile Phase

Here we take the eluotropic series we described for silica and reverse the order. Thus, on silica gel we start with the least polar (hexane) and increase the polarity up to water to carry the more polar components up the plate. On an RP-18 reverse phase, we might start with water – or a water : alcohol, water : acetonitrile, water : propanone, or water : dioxane mixture where water is the major component.

As the polarity of the mixture increases, the water content of the mobile phase should decrease.

In addition, we can modify the selectivity of water : alcohol mixtures by changing from methanol to ethanol to propanol. Thus the mixture ethanol : water (80 : 20 v/v) has proved to be very useful as the first stab at deciding which mobile phase to use. Solvents which can be added to the ethanol water mixtures include tetrahydrofuran, dimethyl formamide and dimethyl sulphoxide.

It should not be thought that you *have* to use water in reverse phase tlc. A very important area is referred to as non-aqueous reverse phase tlc and is based on dichloromethane, hexane and other lipophilic solvents.

Detection

Since the layer contains organic material, it is not possible to use sulphuric acid and the other charring agents. Ultraviolet absorption and fluorescence quenching can be applied as can iodine vapour.

7.2. HIGH PERFORMANCE tlc

High performance thin-layer chromatography provides an improved performance in the sense of increased or improved resolution. Unfortunately such improved resolution has to be paid for.

High performance tlc is not cheap!

You will see that both systems we describe in a moment require a lot of instrumentation. The result is that you rarely use high performance tlc just to see if two components can be separated. We are really dealing with quantitative high performance tlc.

The success of high performance tlc is based on the adsorbent particles which are smaller than those used in the conventional technique. These smaller particles allow very good contact to be made between the sample, the mobile phase and the adsorbent surface.

7.2.1. High Performance tlc on a 5 × 5 cm or a 20 × 20 cm Plate

Π What is the thickness of the adsorbent layer in the plates prepared for conventional tlc?

You will recall that we use 0.25 mm to 1.0 mm thick layers. In high performance tlc, the thickness of the layer is usually 0.10 mm. In one system of high performance tlc, the plate is not noticeably different from the conventional tlc plate and two sizes of plate are available, 5 × 5 and 20 × 20 cm.

Associated with this thinner layer is a smaller and narrower range of particle size (1 to 5 μm) with a very uniform pore size. The small adsorbent particle improves efficiency to such an extent that you can perform the same separation on high performance tlc in 3 to 6 cm of running distance compared to 15 to 20 cm in conventional tlc. This in turn means that the separation is complete in 1/5 or 1/10 of the time needed for conventional tlc and less mobile phase is required.

The smaller adsorbent particles also limit sample size ie if we try to load the same quantity of sample onto the layer as in conventional tlc, it would overload the plate very badly.

Spot Application

Commercial equipment needs to be purchased which will allow the very small volumes of sample solution, 0.5 to 250 nanolitres, to be applied.

Such small volumes allow the placing of the sample in a tiny spot at the origin. Indeed it is suggested that a spot of 0.45 mm in diameter can be produced when the solution is made up in heptane. The same sample would be spread out over a spot of 1.4 mm diameter if it were prepared in propanone.

With such precise spotting, as many as 15 samples can be put on a 5 × 5 cm plate.

Mobile Phase

These are the same as used for conventional tlc but smaller volumes are required which reduces your solvents bill!

Development

Linear development can be performed in the same tanks used for conventional tlc.

Circular development can be performed in Camag's equipment which is shown in Fig. 7.2a.

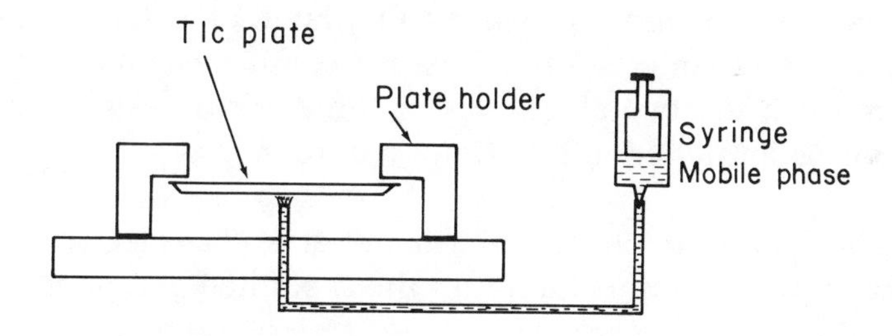

Fig. 7.2a. *Camag U-chamber*

The tlc plate is held upside down in a U-chamber. The mobile phase (up to 10 cm) is held in a syringe which can be pressurised automatically to provide direct a fine jet of liquid onto the silica gel at the centre of the plate. As the solvent moves out in a circle from this point, it carries the sample components with it. Circular high performance tlc is especially effective for components with low R_f values.

In addition to special equipment for the application of the sample and development of the chromatogram, it is usually necessary to use a densitometer as a method of detection to realise the full potential of the technique.

7.2.2. High Performance tlc with the Iatroscan System

A second type of high performance tlc is based on the idea that the silica gel need not be on a flat glass plate but could be on a rod of copper, steel or glass. It has been commercially developed by a Japanese company and marketed as Iatroscan System with the silica gel attached to a glass called a Chromarod. The silica gel is again of a small uniform particle size and regular pore size. Samples are placed on the Chromarod, which is 0.9 mm diameter and 150 mm long, with a syringe. The rod is then placed in a frame which can hold up to 9 rods. The frame is placed in a tlc tank and eluted in the conventional way.

It is in the visualisation step that the Iatroscan is different. The rod, still in its frame, is moved through the flame of a flame ionisation detector. The organic components are combusted and produce a signal between the two electrodes of the detector (Fig. 7.2b). This signal can be amplified and fed to a recorder.

The frame is moved through the flame at a pre-selected speed to ensure complete combustion of all the separated components.

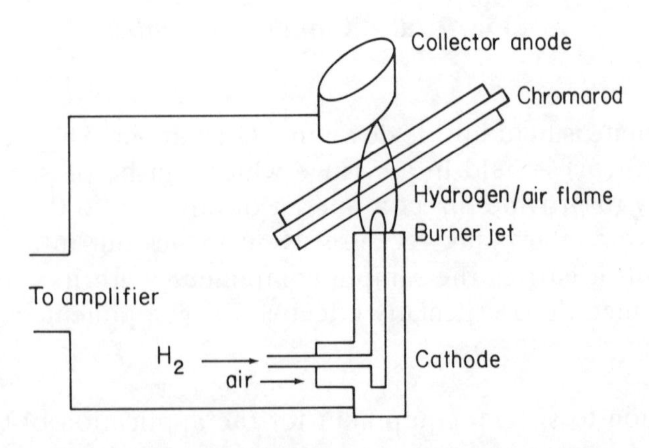

Fig. 7.2b. *Iatroscan system*

The advantage of this system is that the rods can be re-used after they have been cleaned by immersing them in a solution of chromic acid and washing thoroughly with water. They are re-activated by placing them in the frame and passing them through the flame ionisation detector.

The recorder produces a trace which looks like a gas chromatogram. In a typical analysis, a mixture of about equal proportions of cholesteryl acetate and cholesteryl palmitate was analysed (20 replications) with a relative precision of 1.9%.

Either of these methods of high performance tlc is a very sophisticated form of analysis which competes very well with gas or high performance liquid chromatography.

Summary

The use of a modified adsorbent with much smaller particles has increased the efficiency of tlc separations. With these smaller particles less sample is applied to prevent overloading. Visualisation methods must be improved so as to monitor these small quantities. Reverse phase tlc is usually performed with commercially available plates.

Objectives

You should be able to:

- recognise the reversal of the order for an eluotropic series for a reverse phase system;
- explain the differences between high performance and normal tlc.

8 Advantages and Disadvantages of tlc

To illustrate the advantages and disadvantages of tlc let us consider three case studies.

So far we have been piecing together the technique of tlc bit by bit; now let us try to examine the system as a whole.

We have already listed a number of classes of compound which can be analysed by tlc in Fig. 4.1h. In this Section we will describe in more detail just three classes of compound.

Alkaloids

The purine alkaloids present in chocolate, coffee and tea are called theobromine, theophylline and caffeine. They can be separated on silica gel and since they are very polar, they need a very polar mobile phase. B Baehler (*Helv. Chim. Acta*, 1962, 45, 309) used a mixture of ethyl ethanoate : methanol : ethanoic acid (8 : 1 : 1) when the three components eluted in the following order: theobromine R_f 0.36, caffeine R_f 0.41 and theophylline R_f 0.50. The purine alkaloids were removed from the silica gel using a microsublimation technique. An alternative method of visualisation can be achieved by spraying the plate with a solution of alcoholic iodine : potassium iodide followed by 25% hydrochloric acid : 96% ethanol (1 : 1). These three components can also be separated on loose layers of aluminium oxide with a mobile phase of chloroform : butan-1-ol (98 : 2). M Sarsunova and V Schwarz (*Pharmazie*, 1963, 18, 207) found that the alkaloids eluted in the order: theobromine R_f 0.15, theophylline R_f 0.30 and caffeine R_f 0.55.

This changing of the order of elution illustrates one of tlc's major advantages, *viz* the ease with which such changes can be brought about.

Amino Acids

A selection of the many adsorbents which have been used for the separation of amino acids include starch, cellulose, calcium hydroxide, Sephadex and polyacrylonitrile.

M Brenner and A Niederwieser (*Experientia*, 1960, 16, 378) using silica gel G and with two dimensional development showed that tlc could reduce the analysis time to 4–5 hours in contrast to the 2–3 days which were needed to separate amino acids by paper chromatography. Since amino acids are very polar compounds, a very polar mobile phase is needed, eg 96% ethanol : water (7 : 3). With this phase the following R_f values were obtained β-alanine, 0.33, cysteic acid 0.69, leucine 0.03 and tryptophan 0.65.

Ninhydrin is the standard visualisation agent but an alternative is to convert the amino acid into its 1-dimethylaminonaphthalene-sulphonyl derivative which fluoresces in the ultraviolet region of the spectrum.

Here again we see one advantage of tlc in that the visualisation of compounds can be achieved in a number of different ways, each of which has its own particular value.

Oil-soluble Food Dyes

A great deal of work has been devoted to the separation of these dyestuffs which, though polar are not hydrophilic like many of the amino acids with the result that the mobile phases which have been used are more lipophilic. Both silica gel G and aluminium oxide G have been used as stationary phases. In Fig. 8.1, hexane : ethyl acetate (9 : 1) was used with the silica gel G and hexane : ethyl ethanoate (98 : 2) with aluminium oxide. Fig. 8.1 shows again the variation in elution characteristics on changing the adsorbent.

	Silica gel G	Aluminium oxide G
Martius Yellow	0.00	0.00
Butter Yellow	0.68	0.57
Sudan III	0.56	0.41
Sudan G	0.14	0.00

Fig. 8.1. *R_f values of oil soluble food dyes*

These examples illustrate the wide range of compounds which can be analysed by this technique.

In summary then, the advantages of tlc are:

1. It is a very versatile technique that can be applied to almost any class of compound.

2. The separations can be achieved for a minimum of expense, given a good adsorbent and pure solvents.

3. The separations can be achieved in the minimum of time. (Though it has to be admitted that with the shorter hplc columns available now, most workers are aiming for a ten minute analysis time or less.)

4. Tlc is probably the one technique which guarantees success in the separation of an unknown mixture. Being an 'open-column' type of chromatography we can visualise the whole plate. There is nothing hidden as there sometimes is in gas chromatography where some components are not eluted under the conditions used. Thus, each time we try to separate a mixture by tlc we

get a result even if it is only to tell us that we should increase the solvent strength because we can see that all the components are at the origin.

There are some disadvantages:

It can be a messy business especially when you are preparing all your own plates. One researcher suggest that you can always tell where the tlc is done in an establishment by the patina of silica gel which coats every surface in the room.

Whilst tlc can be made as quantitative as many of the other forms of chromatography, such a modification is done by adding a lot of expensive equipment. It can be argued that such a change destroys the major advantages which tlc has, *viz* speed and cheapness. Its major use may therefore remain as a qualitative or semi-quantitative analytical technique.

Self Assessment Questions and Responses

SAQ 1.2a

Choose the best definition for chromatography from the following:

(*i*) Chromatography allows the separation of mixtures by elution with a liquid.

(*ii*) Chromatography requires a competitive equilibrium between the stationary phase, the mobile phase and the sample molecules.

(*iii*) Chromatography permits the preferential sorption of a mixture of components from a gas.

(*iv*) Chromatography is a technique which allows an irreversible attachment between the samples to be separated and an adsorbent.

Response

(*i*) is incorrect because it does not introduce the concept of the third part of the chromatographic system, ie the stationary phase.

(*ii*) is correct because it is sufficiently general to allow for the situation which occurs, in say, gas chromatography as well as tlc since the only difference between the two is that the mobile phase is a gas in one case and a liquid in the other.

(*iii*) is too narrow a definition because it refers only to a gas as the mobile phase.

(*iv*) is incorrect because it describes the phenomenon involved in charcoal purification.

****

SAQ 2.4a

Imagine that you are working in the County Analyst's Laboratory and your boss brings in a sample of methyl esters which he has prepared from the triglycerides present in a table margarine. He asks you to obtain 10 mg samples of both saturated and unsaturated methyl esters so that you can check that the label on the margarine packet is correct when it says 'High in Polyunsaturates'.

Which of the following tlc plates would you prepare?

(*i*) 0.25 mm layer thickness of silver nitrate/silica gel.

(*ii*) 1.00 mm layer thickness of silica gel.

(*iii*) 0.25 mm layer thickness of cellulose.

(*iv*) 1.00 mm layer thickness of silver nitrate/silica gel.

(*v*) 0.25 mm layer thickness of silica gel.

Response

(*i*) Whilst it is possible to separate for preparative purposes on a 0.25 mm layer, you have to reduce the quantity of sample so you would need to use two to three times as many plates to separate 10 mg of methyl esters.

(*ii*) This is the best layer thickness for preparative purposes but silica gel separates according to the number of polar functional groups in a molecule. In this margarine sample, both saturated and unsaturated methyl esters have one polar functional group, ie the $-COOCH_3$ group.

(*iii*) Cellulose is an adsorbent which is best used, like paper, for water soluble components. Methyl esters from fats are very water insoluble (hydrophobic) and so do not separate on cellulose according to the number of double bonds.

(*iv*) Correct Response. This layer thickness is better for preparative purposes and the silver ions on the silica gel interact differently with monounsaturated compounds compared to diunsaturated compounds. Thus saturated methyl esters can be separated from methyl polyunsaturated esters.

(*v*) This is too thin a layer for preparative purposes and the silica gel will not separate mixtures according to the number of double bonds.

****

SAQ 3.2a

In eluting samples on silica gel, is the first solvent of each pair more or less polar than the second?

Ethanol is more/less polar than benzene.

Trichloroethylene is more/less polar than cyclohexane.

Ethoxyethane is more/less polar than methanol.

Propanol is more/less polar than ethanol.

Response

Solvents at the beginning of the list which precedes this question are more effective than those at the end, and a list like this, which places solvents in order of eluting power is called an *eluotropic series*.

Ethanol is more polar than benzene.

Trichloroethylene is more polar than cyclohexane.

Ethoxyethane is less polar than methanol.

Propanol is less polar than ethanol.

****

SAQ 3.2b

In the third trial above if you were trying to find a suitable mobile phase by mixing dichloromethane with another solvent, which of the following would you choose:

1. pentane

2. trichloromethane

3. ethanol?

Response

(*i*) If you mix dichloromethane with pentane, you produce a solvent with an ϵ^0 value between 0.42 and 0.00. This will result in sample components having R_f values between 0.1 and 0.9. They will not be at the origin or at the solvent front.

(*ii*) Trichloromethane has an ϵ^0 value of 0.40 so that it will not depress the ϵ^0 value of dichloromethane very much. The result will be an R_f values of about 0.95, ie the sample components are very close to the solvent front.

(*iii*) Ethanol is a very polar solvent (ϵ^0 value of 0.88) so that the sample components will stay very close to the solvent front.

****

SAQ 3.2c

Which of the following concentrations of acetonitrile in pentane would give a solvent strength (ϵ^0 value) of 0.45?

(*i*) 1.0%,

(*ii*) 30%,

(*iii*) 40%,

(*iv*) 80%.

Response

(*i*) You are using the bottom line which represents methanol in pentane.

(*ii*) This is correct.

(*iii*) This would give an ϵ^0 value of 0.46.

(*iv*) This would give an ϵ^0 value of 0.48.

****

SAQ 3.2d

Which of the following proportions of acetonitrile in dichloromethane would give a solvent strength of 0.35?

(*i*) 1.0%,

(*ii*) 3.0%,

(*iii*) 10%,

(*iv*) 100%.

Response

3% acetonitrile in dichloromethane

****

SAQ 3.2e

What is the ϵ^{o} value of a solution containing 10% methanol in ethoxyethane?

Response

$\epsilon^{o} = 0.60$

****

SAQ 3.2f

Two components A and B in a mixture have almost identical R_f values of 0.02 and 0.03 on silica gel with a mobile phase of 2-chloropropane : pentane = 8 : 92.

Which of the following mobile phases would you use to try to increase the separation of the components A and B?

(*i*) pentane,

(*ii*) methyl ethanoate,

(*iii*) acetonitrile : pentane 20 : 80,

(*iv*) methanol : dichloromethane 1.5 : 98.5.

Response

(*i*) pentane has an ϵ^0 value of 0.00 so that the interaction between pentane and compounds A and B will be weaker than for 2-chloropropane : pentane (8 : 92). So compounds A and B will be adsorbed still more firmly, and the R_f values might be as low as 0.00.

(*ii*) Methyl ethanoate has an ϵ^0 value of 0.60. This is at the other extreme from pentane. Although we cannot be certain, it is likely that such a solvent will compete too well with compounds A and B for the active sites on the adsorbent so it will cause them to elute far up the plate with large R_f value, perhaps even at the solvent front.

(*iii*) This solvent mixture has an ϵ^0 value of 0.43 which represents a large increase compared to 2-chloropropane : pentane (8 : 92), but hopefully not so large as methyl ethanoate. We could hope that this solvent mixture would move the two compounds A and B part way up the plate into the preferred R_f range of 0.3–0.7.

(*iv*) This solvent mixture also has an ϵ^o value of 0.43. If acetonitrile : pentane did not give as good a selectivity as you might wish, ie the two R_f values were not too far apart, then this mixture might give a better separation, depending on the chemical nature of compounds A and B.

****

SAQ 4.1a

In relation to the separation of an aliphatic ketone from an aliphatic alcohol on silica gel, which one of the following paragraphs, (*i*)–(*iv*), is correct?

(*i*) The aliphatic alcohol R—OH will have an R_f value of 0.2 compared to a value of 0.5 for the aliphatic ketone R—CO—R because the alcohol interacts only by London Dispersion forces.

(*ii*) The aliphatic ketone will have an R_f value of 0.2 compared to a value of 0.5 for the aliphatic alcohol because the ketone can interact by hydrogen bonding as a proton donor.

(*iii*) The aliphatic alcohol will have an R_f value of 0.2 compared to a value of 0.5 for the aliphatic ketone because the alcohol interacts by hydrogen bonding as a proton acceptor and as a proton donor. ⟶

SAQ 4.1a (cont.)

(*iv*) The aliphatic ketone will have an R_f value of 0.2 compared to a value of 0.5 for the aliphatic alcohol because the ketone will interact by chemisorption.

Response

(*i*) is incorrect. Whilst all aliphatic molecules will interact by means of the London Dispersion forces, both the ketone and alcohol will interact to a similar extent if the R-groups are not too dissimilar.

(*ii*) is incorrect. The ketone will not be retained as strongly as the alcohol so the R_f values are wrong. In addition, the ketone cannot act as a proton donor because it does not have a hydrogen attached to a hetero-atom. The ketone can act as a proton acceptor.

(*iii*) is correct. The alcohol will have the smaller R_f value and whilst the ketone will interact by London Dispersion forces on the CH_2 and CH_3 part of the chain and by hydrogen bonding as a proton acceptor, the alcohol interacts by both these forces and by hydrogen bonding as a proton donor.

(*iv*) is incorrect. The major groups which can interact by chemisorption are amines and carboxylic acids. The ketone will *not* form covalent bonds to silica gel.

****

SAQ 5.1a

How can we select a solvent with confidence from those in which the samples dissolve and be sure that we minimise interaction of the selected solvent with the adsorbent? Choose from the following ideas:

(*i*) Select a solvent of minimum boiling point,

(*ii*) Select a solvent high in the eluotropic series,

(*iii*) Select a less reactive solvent,

(*iv*) Select a solvent low in the eluotropic series.

Response

(*i*) It is important that the solvent has a low boiling point but in order to decrease attraction to the adsorbent, (*iv*) is also important.

(*ii*) No. You would not do this as a solvent high in the series competes too strongly with sample molecules for active sites.

(*iii*) Any solvent should be unreactive to samples and adsorbents in a given separation, so 'less' reactive is not an important criterion.

(*iv*) Yes, you should choose a solvent with a low ϵ^0 value and preferably with a low boiling point also.

****

SAQ 5.1b

Given a choice of four solvents for a sample from the following:

cyclohexane, pyridine, chlorobenzene, water,

which would you choose as your ideal spotting solvent?

Response

The ϵ^0 values point to cyclohexane. Boiling points also suggest cyclohexane.

A word of caution – ethoxyethane seems to meet many of the requirements of a solvent but it is, if anything, too volatile. It is difficult to keep ether solutions without losing the solvent. So the concentration varies from application to application on the plate. It is also difficult to transfer ether solutions with microsyringes or micropipettes without the ether boiling and the solution being ejected before you get to the plate.

****

SAQ 5.1c

Given a 10 cm^3 solution containing 20.0 g of a sample and diluting your solution by one hundred each time, how many dilutions are required to achieve a concentration of 0.2 g dm^{-3}?

Response

20.0 g in 10.0 cm^3 is equal to 2,000 mg in 1 cm^3.

Therefore, diluting 1.0 cm^3 to 100 cm^3, will give a solution 2,000 mg per 100 cm^3 containing 20 mg cm^{-3} after *one* dilution.

Taking 1 cm^3 of this solution and diluting it to 100 cm^3, will give a solution containing 20.0 mg per 100 cm^3 containing 0.2 mg cm^{-3} after *two* dilutions.

This corresponds 200 μg cm^{-3} or 200 μg per 1000 μl.

This is the required concentration 0.2 g dm^{-3}.

****

SAQ 5.1d

Resolution can also be improved by increasing X, the distance between spot centres. By reference to previous Sections where you think it is appropriate, is this improved resolution influenced by:

(*i*) the choice of adsorbent,

(*ii*) the shape of the chromatographic tank,

(*iii*) the nature of the mobile phase,

(*iv*) the nature of the spotting solvent?

Response

The correct answers are (*i*) and (*iii*). Selectivity and hence resolution can often be improved by altering either the stationary phase or, more easily, the composition of the mobile phase. The shape of the tank and the nature of the spotting solvent are irrelevant in this context.

****

SAQ 5.1e

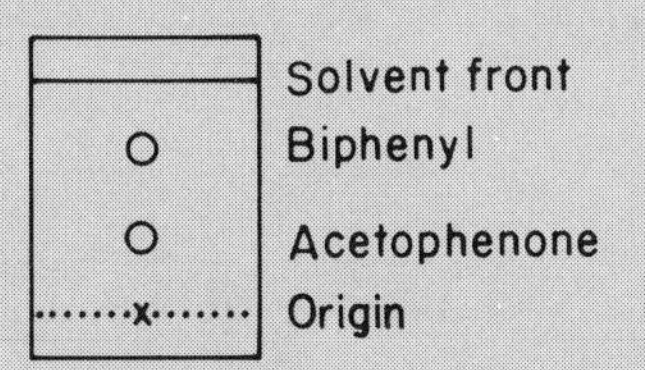

Fig. 5.1f. *Separation of biphenyl and acetophenone*

From the above diagram for the tlc separation of acetophenone and biphenyl,

(*i*) Show that the resolution is equal to 2.5.

(*ii*) Choose from the following responses to explain how resolution could be increased.

(*a*) We could use a thinner layer on the tlc plate with the same amount of sample.

(*b*) The sample could be applied to the plate in a solvent such as methanol without allowing the methanol to dry between each application. ⟶

SAQ 5.1e (cont.)

(*c*) We could choose a mobile phase with the same ϵ^0 value but with a different composition.

(*d*) We could put less sample on the tlc plate.

(*e*) We could choose a mobile phase with the same ϵ^0 value but with a different composition and also put less sample on the plate.

Response

(*i*) Resolution $= \dfrac{\text{Distance between centres of spots}}{\text{Average diameter of spots}}$

$$= \frac{0.5}{0.5(0.2 + 0.2)} = \frac{0.5}{0.5(0.4)}$$

$$= 2.5$$

(*ii*)

(*a*) If we go to a thinner layer we may overload the adsorbent and so we may get more overlap of the spots, so resolution would decrease.

(*b*) This is the recipe for producing rings of sample where the spot size gets gradually larger.

(*c*) This is a partially correct response.

(*d*) This is also a partially correct response.

(*e*) This is the correct response.

****

SAQ 5.3a

Indicate whether each of the following statements is TRUE or FALSE.

(*i*) In visualising an amino acid separation by spraying the plate with ninhydrin, a research worker was using a destructive technique.

TRUE / FALSE?

(*ii*) After separating fatty acids on a silica gel GF, a research worker examined the spots under an ultra violet lamp. A chemical method of visualisation was being used.

TRUE / FALSE?

(*iii*) In preparative tlc a worker says destructive visualisation techniques are preferable to non-destructive techniques.

TRUE / FALSE?

(*iv*) You can use water as an appropriate non-destructive visualising agent with esters.

TRUE / FALSE?

(*v*) Inorganic compounds can be visualised by using sulphuric acid as a destructive agent.

TRUE / FALSE?

(*vi*) Ultraviolet radiation can be used as a non-destructive visualisation technique only with coloured substances.

TRUE / FALSE?

Response

(*i*) TRUE. The ninhydrin reacts with the amino acid to give a coloured derivative which appears in the visible region of the spectrum.

(*ii*) FALSE. Ultraviolet visualisation depend on the physical interaction of the ultraviolet radiation with the phosphor in the silica gel GF not a chemical reaction.

(*iii*) FALSE. In preparative tlc you wish to recover the product so you *must* visualise non-destructively.

(*iv*) FALSE. Whilst in most cases water is a non-destructive agent, some esters are hydrolysed.

(*v*) FALSE. In most inorganic substances there is no carbon for the sulphuric acid to char. Some inorganic substances *may* give colours with sulphuric acid.

(*vi*) FALSE. The use of ultraviolet radiation is an important non-destructive method but it can be used with substances other than coloured ones.

****

SAQ 6.1a

Which of the following would help to ensure a saturated atmosphere in a tank?

(*i*) Leaving the tank for 5 minutes or less before developing the plate (refer our worker on day two).

(*ii*) Leaving the tank for an hour or more before developing the plate (refer our worker on day one). ⟶

SAQ 6.1a (cont.)

(*iii*) Lining the tlc tank with filter paper.

(*iv*) Using a larger tlc tank with respect to the size of the plate.

(*v*) Using as small a tank as possible with respect to the size of the plate.

Response

(*i*) 5 minutes or less would be too short a time to allow vapour to saturate the air and equilibrate.

(*ii*) An hour or more would be much more reasonable and of course the larger the tank the longer the time. If the tank is lined with filter paper then 15 to 20 minutes is quite sufficient.

(*iii*) Yes – filter paper does help to ensure saturation of the air with vapour (see (*ii*) above).

(*iv*) No – in a large tank it is more difficult to ensure saturation and a large tank does not assist the chromatographic process in any way.

(*v*) Yes – the smaller the volume of tank that requires saturation the better and this reduces the time required for saturation (see (*ii*)).

****

SAQ 6.1b

What change in ϵ^0 value would you expect to find if dichloromethane contained 1% methanol, compared to pure dichloromethane?

Response

Referring to Fig. 3.1b it shows that 0.32 would change to 0.45. Such a change would produce a large alteration in R_f value. In the case of trichloromethane, it normally contains 1.0–1.5% ethanol as stabilizer so its ϵ^0 value is changed by this addition. In considering other research workers' R_f values, it would be important to check whether they used trichloromethane straight from the bottle or whether they distilled it. Very different R_f values could result.

****

SAQ 6.1c

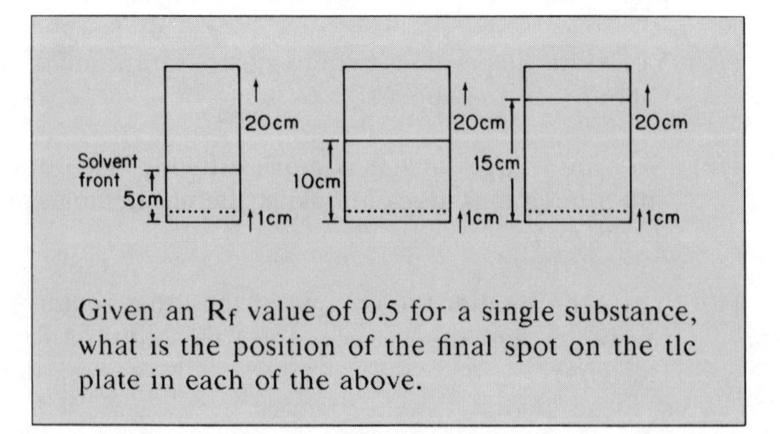

Given an R_f value of 0.5 for a single substance, what is the position of the final spot on the tlc plate in each of the above.

Response

2 cm above origin, 4.5 cm above origin; 7 cm above origin.

****

Units of Measurement

For historic reasons a number of different units of measurement have evolved to express quantity of the same thing. In the 1960s, many international scientific bodies recommended the standardisation of names and symbols and the adoption universally of a coherent set of units—the SI units (Système Internationale d'Unités)—based on the definition of five basic units: metre (m); kilogram (kg); second (s); ampere (A); mole (mol); and candela (cd).

The earlier literature references and some of the older text books, naturally use the older units. Even now many practicing scientists have not adopted the SI unit as their working unit. It is therefore necessary to know of the older units and be able to interconvert with SI units.

In this series of texts SI units are used as standard practice. However in areas of activity where their use has not become general practice, eg biologically based laboratories, the earlier defined units are used. This is explained in the study guide to each unit.

Table 1 shows some symbols and abbreviations commonly used in analytical chemistry. Table 2 shows some of the alternative methods for expressing the values of physical quantities and the relationship to the value in SI units.

More details and definition of other units may be found in the *Manual of Symbols and Terminology for Physicochemical Quantities and Units*, Whiffen, 1979, Pergamon Press.

Table 1 *Symbols and Abbreviations Commonly used in Analytical Chemistry*

Symbol	Meaning
Å	Angstrom
A_r(X)	relative atomic mass of X
A	ampere
E or *U*	energy
G	Gibbs free energy (function)
H	enthalpy
J	joule
K	kelvin (273.15 + *t* °C)
K	equilibrium constant (with subscripts p, c, therm etc.)
K_a,K_b	acid and base ionisation constants
M_r(X)	relative molecular mass of X
N	newton (SI unit of force)
P	total pressure
s	standard deviation
T	temperature/K
V	volume
V	volt (J A^{-1} s^{-1})
a, *a*(A)	activity, activity of A
c	concentration/ mol dm^{-3}
e	electron
g	gramme
i	current
s	second
t	temperature / °C
bp	boiling point
fp	freezing point
mp	melting point
$\approx$	approximately equal to
$<$	less than
$>$	greater than
e, exp(*x*)	exponential of *x*
ln *x*	natural logarithm of *x*; ln *x* = 2.303 log *x*
log *x*	common logarithm of *x* to base 10

Table 2 *Alternative Methods of Expressing Various Physical Quantities*

1. **Mass (SI unit : kg)**

$$\mathrm{g} = 10^{-3}\ \mathrm{kg}$$
$$\mathrm{mg} = 10^{-3}\ \mathrm{g} = 10^{-6}\ \mathrm{kg}$$
$$\mu\mathrm{g} = 10^{-6}\ \mathrm{g} = 10^{-9}\ \mathrm{kg}$$

2. **Length (SI unit : m)**

$$\mathrm{cm} = 10^{-2}\ \mathrm{m}$$
$$\text{Å} = 10^{-10}\ \mathrm{m}$$
$$\mathrm{nm} = 10^{-9}\ \mathrm{m} = 10\text{Å}$$
$$\mathrm{pm} = 10^{-12}\ \mathrm{m} = 10^{-2}\ \text{Å}$$

3. **Volume (SI unit : m^3)**

$$\mathrm{l} = \mathrm{dm}^3 = 10^{-3}\ \mathrm{m}^3$$
$$\mathrm{ml} = \mathrm{cm}^3 = 10^{-6}\ \mathrm{m}^3$$
$$\mu\mathrm{l} = 10^{-3}\ \mathrm{cm}^3$$

4. **Concentration (SI units : mol m^{-3})**

$$\mathrm{M} = \mathrm{mol\ l^{-1}} = \mathrm{mol\ dm^{-3}} = 10^3\ \mathrm{mol\ m^{-3}}$$
$$\mathrm{mg\ l^{-1}} = \mu\mathrm{g\ cm^{-3}} = \mathrm{ppm} = 10^{-3}\ \mathrm{g\ dm^{-3}}$$
$$\mu\mathrm{g\ g^{-1}} = \mathrm{ppm} = 10^{-6}\ \mathrm{g\ g^{-1}}$$
$$\mathrm{ng\ cm^{-3}} = 10^{-6}\ \mathrm{g\ dm^{-3}}$$
$$\mathrm{ng\ dm^{-3}} = \mathrm{pg\ cm^{-3}}$$
$$\mathrm{pg\ g^{-1}} = \mathrm{ppb} = 10^{-12}\ \mathrm{g\ g^{-1}}$$
$$\mathrm{mg\%} = 10^{-2}\ \mathrm{g\ dm^{-3}}$$
$$\mu\mathrm{g\%} = 10^{-5}\ \mathrm{g\ dm^{-3}}$$

5. **Pressure (SI unit : N m^{-2} = kg m^{-1} s^{-2})**

$$\mathrm{Pa} = \mathrm{Nm^{-2}}$$
$$\mathrm{atmos} = 101\ 325\ \mathrm{N\ m^{-2}}$$
$$\mathrm{bar} = 10^5\ \mathrm{N\ m^{-2}}$$
$$\mathrm{torr} = \mathrm{mmHg} = 133.322\ \mathrm{N\ m^{-2}}$$

6. **Energy (SI unit : J = kg m^2 s^{-2})**

$$\mathrm{cal} = 4.184\ \mathrm{J}$$
$$\mathrm{erg} = 10^{-7}\ \mathrm{J}$$
$$\mathrm{eV} = 1.602 \times 10^{-19}\ \mathrm{J}$$

Table 3 *Prefixes for SI Units*

Fraction	Prefix	Symbol
10^{-1}	deci	d
10^{-2}	centi	c
10^{-3}	milli	m
10^{-6}	micro	μ
10^{-9}	nano	n
10^{-12}	pico	p
10^{-15}	femto	f
10^{-18}	atto	a

Multiple	Prefix	Symbol
10	deka	da
10^{2}	hecto	h
10^{3}	kilo	k
10^{6}	mega	M
10^{9}	giga	G
10^{12}	tera	T
10^{15}	peta	P
10^{18}	exa	E

Table 4 *Recommended Values of Physical Constants*

Physical constant	Symbol	Value
acceleration due to gravity	g	$9.81\ m\ s^{-2}$
Avogadro constant	N_A	$6.022\ 05 \times 10^{23}\ mol^{-1}$
Boltzmann constant	k	$1.380\ 66 \times 10^{-23}\ J\ K^{-1}$
charge to mass ratio	e/m	$1.758\ 796 \times 10^{11}\ C\ kg^{-1}$
electronic charge	e	$1.602\ 19 \times 10^{-19}\ C$
Faraday constant	F	$9.648\ 46 \times 10^{4}\ C\ mol^{-1}$
gas constant	R	$8.314\ J\ K^{-1}\ mol^{-1}$
'ice-point' temperature	T_{ice}	273.150 K exactly
molar volume of ideal gas (stp)	V_m	$2.241\ 38 \times 10^{-2}\ m^3\ mol^{-1}$
permittivity of a vacuum	ϵ_o	$8.854\ 188 \times 10^{-12}\ kg^{-1}\ m^{-3}\ s^4\ A^2\ (F\ m^{-1})$
Planck constant	h	$6.626\ 2 \times 10^{-34}\ J\ s$
standard atmosphere pressure	p	$101\ 325\ N\ m^{-2}$ exactly
atomic mass unit	m_u	$1.660\ 566 \times 10^{-27}\ kg$
speed of light in a vacuum	c	$2.997\ 925 \times 10^{8}\ m\ s^{-1}$